London Mathematical Society
Lecture Note Series

This series publishes the records of lectures and seminars on advanced topics in mathematics held at universities throughout the world. For the most part, these are at postgraduate level either presenting new material or describing older material in a new way. Exceptionally, topics at the undergraduate level may be published if the treatment is sufficiently original.

Prospective authors should contact the editor in the first instance.

Already published in this series

1. General cohomology theory and K-theory. Peter Hilton
2. Numerical ranges of operators on normed spaces and of elements of normed algebras. F. F. Bonsall and J. Duncan
3. Convex polytopes and the upper bound conjecture. P. McMullen and G. C. Shephard
4. Algebraic topology: A student's guide. J. F. Adams
5. Commutative algebra. J. T. Knight
6. Finite groups of automorphisms. Norman Biggs
7. Introduction to combinatory logic. J. R. Hindley, B. Lercher and J. P. Seldin
8. Integration and harmonic analysis on compact groups. R. E. Edwards
9. Elliptic functions and elliptic curves. Patrick du Val
10. Numerical ranges II. F. F. Bonsall and J. Duncan
11. New developments in topology. G. Segal (ed.)
12. Proceedings of the Symposium in Complex Analysis, Canterbury, 1973. J. Clunie and W. K. Hayman (eds.)
13. Combinatorics, Proceedings of the British Combinatorial Conference, 1973. T. P. McDonough and V. C. Mavron (eds.)

London Mathematical Society Lecture Note Series. 14

Analytic Theory of Abelian Varieties

H. P. F. SWINNERTON-DYER

Cambridge · At the University Press · 1974

CAMBRIDGE UNIVERSITY PRESS
Cambridge, New York, Melbourne, Madrid, Cape Town, Singapore, São Paulo

Cambridge University Press
The Edinburgh Building, Cambridge CB2 8RU, UK

Published in the United States of America by Cambridge University Press, New York

www.cambridge.org
Information on this title: www.cambridge.org/9780521205269

First published 1974
Re-issued in this digitally printed version 2008

A catalogue record for this publication is available from the British Library

Library of Congress Catalogue Card Number: 74-77835

ISBN 978-0-521-20526-9 paperback

Contents

Introduction

For fixed $n > 0$ let $\mathbf{C}^n$ denote n-dimensional complex space and $\mathbf{R}^{2n}$ the underlying 2n-dimensional real space. Let Λ be a lattice in $\mathbf{R}^{2n}$ - that is, a free abelian group on $2n$ generators which spans $\mathbf{R}^{2n}$; thus with the induced topology Λ is discrete and $T = \mathbf{R}^{2n}/\Lambda$ is compact. T has a natural structure as a complex manifold, which we recognize by writing it as $T = \mathbf{C}^n/\Lambda$, and with this structure it is called a complex torus. The most interesting case is when there are sufficiently many meromorphic functions on T, in a sense that will be made precise in Chapter II; in this case T is called an abelian manifold, and is usually denoted by A. Thus from one point of view the study of abelian manifolds is essentially the study of meromorphic functions of n complex variables having 2n independent periods. Thus it forms a natural generalization of the theory of elliptic functions - that is, of doubly periodic functions of one complex variable, Historically, the other parent of the subject was the study of compact Riemann surfaces; indeed the term 'abelian manifolds' comes from the connection with Abel's theorem.

A compact Riemann surface is just another way of describing a non-singular algebraic curve; and this already gives a connection between algebraic geometry and abelian manifolds. This connection was much strengthened by Lefschetz and the great Italian geometers, who showed that abelian manifolds are an important tool in the problem of classifying non-singular varieties. Conversely any abelian manifold can be embedded in projective space as a variety in the sense of algebraic geometry; when we wish to emphasize this point of view, we speak of an abelian variety instead of an abelian manifold. The study of abelian varieties by purely geometric methods, valid over fields of arbitrary characteristic, was initiated by Weil; see [14], [15] and Lang [5]. (Numbers in square brackets refer to the list of references at the end of this volume.) An up-to-date account may be found in Mumford [8]. More recently Shimura and others have shown that the theory of abelian manifolds has important application to algebraic number theory; for Shimura's work see [9], [10] and a series of papers published during the 1960s, largely in Annals of Mathematics.

The main object of this book is to give an account of the standard theory of abelian manifolds which presupposes not much more than a basic complex variable course. It contains all the material on abelian manifolds which is needed for the applications to algebraic geometry and to number theory; indeed it is based on a course of lectures delivered in Cambridge which was designed to lead on to an exposition of some of Shimura's work. But it does not contain an exposition of either of these applications. I have included some geometrical results, and they do presuppose a knowledge of algebraic geometry; the reader who lacks this may omit them without inconvenience. Also §10, on the structure of the ring of endomorphisms of an abelian variety, requires some knowledge of the theory of algebras and of algebraic number theory; however the necessary results from the theory of algebras are stated without proof at the beginning of §10.

The book starts with two sections closely connected with the main theme though not essential to it. In §1 there is a survey of the theory of compact Riemann surfaces - without proofs of the key theorems because to include proofs would take so much space that it would unbalance the book. Proofs may be found for example in Gunning [4]. In §2 there is a brief account, with proofs, of the theory of elliptic functions. This is the special case $n = 1$ of the theory, but it is untypical for two reasons. Because one can give a simple description of divisors when $n = 1$, there are many explicit formulae in that case which cannot be generalized; also the Riemann form, which plays a major part in the general theory, appears in so trivial a form when $n = 1$ that its presence goes unnoticed. The rest of Chapter I contains some general results on functions of several complex variables which are needed later. Chapter II is concerned with necessary and sufficient conditions on Λ for $T = \mathbf{C}^n/\Lambda$ to admit non-constant meromorphic functions, or to be an abelian manifold, and with the construction of those functions. It also contains the immediate consequences of these results, in particular theorems about the projective embedding of abelian manifolds. Chapter III contains the standard theory of abelian manifolds themselves, in particular the study of the ring of endomorphisms of an abelian manifold and the matrix representations of these endomorphisms, and the duality theory which essentially describes the group of divisors on an abelian manifold. Many of the results in this Chapter can be stated in purely algebro-geometric terms, even though the proofs which have been given for them are analytic. I have therefore given in an Appendix a brief account without proofs of the geometric theory of abelian varieties over a field of arbitrary characteristic, to show how far the analytic theory remains valid.

The most elegant account of the analytic theory is that given in Chapter VI of Weil [17]; but that account was written to show how Hodge theory can be applied to a particularly simple kind of complex manifold and it therefore assumes a great deal of background knowledge. The same is true of the account in Mumford [8], Chapter I. Two books which assume no more knowledge than this one are Conforto [2] and Lang [6]. Conforto's treatment is very old-fashioned; in particular a substantial part of his book is taken up by Poincaré's original proof of the key existence theorem, that to every positive divisor on T corresponds a theta-function on $\mathbf{C}^n$, whereas the proof given here, based on Weil [16], only takes a few pages. Lang's book is very similar in spirit to this one, though perhaps more modern; but I believe there is enough difference in the material covered to justify publishing this also. Much of the theory can also be found in Siegel [11].

Nothing in this book is original, except perhaps the errors. I have therefore only ascribed a result to someone if it is generally known by his name.

I would like to express my gratitude to André Weil, Serge Lang and S. J. Patterson for the help which they have given me at various levels; without them this book would never have been written.

NOTATION

As is usual, **C**, **Q**, **R** and **Z** denote respectively the complex numbers, the rationals, the reals and the rational integers; but Q is sometimes also used to denote a quadratic form. From the beginning of

Chapter II onwards, $V = \mathbf{C}^n$ is a complex vector space of dimension n and Λ is a lattice in V; the complex torus V/Λ is denoted by T in general, and by A when it is known to be an abelian manifold. In the Appendix, A is an abelian variety of dimension n defined over an arbitrary field k, not necessarily of characteristic zero. Both

$$z = (z_1, \ldots, z_n) \quad \text{and} \quad w = (w_1, \ldots, w_n)$$

usually denote points of V and their coordinates with respect to a fixed base for V; they may also denote the corresponding points on T or A. Similarly λ denotes an element of Λ; but note that $\lambda_1, \ldots, \lambda_{2n}$ is a base for Λ, rather than a set of coordinates for λ. Finally E and H denote respectively the alternating and the Hermitian version of a Riemann form on $V \times V$ with respect to Λ. By convention, a Hermitian form will be $\mathbf{C}$-linear in its first argument and $\mathbf{C}$-antilinear in its second; the reader is warned that Weil [17] adopts the opposite convention.

Since the distinction between Theorems and Lemmas is purely subjective, they share a common system of numbering; however this principle has not been extended to Corollaries.

I · Background

1. Compact Riemann surfaces and algebraic curves

A Riemann surface is a topological space which in a neighbourhood of any point looks like the complex plane. (Some writers also require a Riemann surface to be connected.) More precisely, a Riemann surface is a Hausdorff space S which is endowed with a complex one-dimensional structure in the following way. For each point P of S we are given an open neighbourhood N of P and a homeomorphism ϕ from N to a disc $|z - \phi(P)| < c$ in the complex z-plane for some $c > 0$. These homeomorphisms satisfy the following consistency condition. Let P_1, P_2 be any two points of S such that their corresponding neighbourhoods N_1, N_2 overlap; then $\phi_1 \circ \phi_2^{-1}$ is holomorphic on $\phi_2(N_1 \cap N_2)$. If $f(z)$ is a function which is holomorphic and has non-zero derivative at $z = \phi(P)$, then we can replace ϕ by $f \circ \phi$ (with corresponding changes in N and c) without changing the complex structure of S. We call ϕ a local variable at P; clearly it is then a local variable at each point of N.

We can now transfer all the standard terminology of the theory of functions of a complex variable from the complex plane to Riemann surfaces. For example, a function ψ defined in a neighbourhood of P is holomorphic at P if $\psi \circ \phi^{-1}$ is holomorphic at $z = \phi(P)$, where ϕ is a local variable at P; the consistency condition is just what is needed to ensure that ψ is then holomorphic in a neighbourhood of P. Again, ψ has a zero of order n at P if $\psi \circ \phi^{-1}$ has a zero of order n at $z = \phi(P)$, and similarly for poles. A differential ω defined in a neighbourhood of P can be written in the form $\psi d\phi$; it is said to be holomorphic or meromorphic at P, or to have a zero or pole of order n at P, if ψ has the corresponding property. It is easy to see that none of these properties depends on the choice of the local variable ϕ. It must be remembered that a residue is associated with a differential rather than with a function. If $\omega = \psi d\phi$ has at worst an isolated singularity at P, then its residue at P is $(2\pi i)^{-1}\int\omega$ taken round a closed contour Γ which lies in N, goes once round P anticlockwise under the orientation induced in N by ϕ, and contains no singularity of ψ inside or on it except perhaps P itself. It follows that the residue is equal to the coefficient of $1/(\phi - \phi(P))$ in the Laurent series expansion of ψ in terms of $\phi - \phi(P)$, and also that it is independent of the choice of ϕ.

In general one is primarily interested in those functions and differentials on a Riemann surface which are everywhere meromorphic; and in what follows we shall confine ourselves to them. A differential is said to be of the <u>first kind</u> on a Riemann surface S if it is holomorphic everywhere on S, of the <u>second kind</u> if it is meromorphic on S and has residue zero at every pole, and of the <u>third kind</u> if it is meromorphic on S; differentials of any given kind form a **C**-vector space. A differential is called <u>exact</u> if it is of the form $d\psi$ where ψ is meromorphic on S; clearly an exact differential is of the second kind, but not every differential of the second kind is exact.

We can now translate Cauchy's theorem into the language of Riemann surfaces:

Theorem 1. <u>Let Γ be a contour on S, not necessarily closed, and let ω be a meromorphic differential on S. Suppose that Γ varies continuously, with its endpoints if any remaining fixed. Then $\int_\Gamma \omega$ remains constant as long as Γ does not move across a singularity of ω, and it changes by $2\pi i$ times the residue at a singularity whenever Γ moves across that singularity.</u>

In particular, if Γ is a closed curve and ω is of the first or second kind, the value of $\int_\Gamma \omega$ depends only on the homology class of Γ; this value is called the <u>period</u> of ω with respect to Γ. It is easy to see that ω is exact if and only if all its periods are equal to zero.

Henceforth we shall also assume that S is compact, and usually that it is connected. (A compact Riemann surface is the disjoint union of finitely many compact connected Riemann surfaces, so that results in the general case follow at once from those in the connected case; but they are often more complicated to state in the general case.) At each point of S there is a canonical local orientation, obtained by means of

ϕ^{-1} from the canonical orientation of the complex plane. These orientations are compatible, so S itself is oriented. Since it is compact it can be triangulated and the triangulation is finite; topologically S is just a sphere with g handles, for some $g \geq 0$, and its first homology group $H_1(S, \mathbf{Z})$ is a free abelian group on 2g generators.

Theorem 2. (i) *If* ω *is a meromorphic differential on a compact Riemann surface* S *then* ω *has only finitely many poles on* S *and the sum of the residues of* ω *at these poles is* 0.

(ii) *If* ψ *is a non-constant meromorphic function on a compact Riemann surface* S *then* ψ *takes every value the same number of times (allowing for multiplicities). In particular* ψ *has as many poles as zeros.*

Proof. If ω had infinitely many poles these would have a point of accumulation on S, by compactness, and ω would not be meromorphic at that point. Now triangulate S, integrate ω round each triangle and sum the results; (i) follows at once since in the sum ω has been integrated twice along each side of each triangle, once in each direction. Applying these results to the special case $\omega = d\psi/\psi$ we find that ψ has as many poles as zeros, and writing $\psi - c$ for ψ we find that ψ takes the value c as many times as it has poles. This completes the proof of the Theorem. The number of times ψ takes each value is called the *valence* of ψ. Constant functions are deemed to have valence 0.

Lemma 3. *Let* θ, ψ *be non-constant meromorphic functions on a compact Riemann surface* S, *of valences* m, n *respectively. Then there is a polynomial* $F(X, Y)$ *of degrees at most* n *in* X *and* m *in* Y *such that* $F(\theta, \phi) = 0$. *If moreover* S *is connected then* F *can be chosen to be irreducible.*

Proof. Choose a complex number c which is not a value taken by ψ at any of the zeros or poles of θ, and let $P_1, \ldots, P_n$ be the points of S at which $\psi = c$. For any $r \geq 0$ the differential $\omega = \theta^r d\psi/(\psi - c)$ has a simple pole with residue $\theta^r(P_\nu)$ at each P_ν; its other poles are at the poles of θ and ψ, and at each of these the residue is a rational function of c. It now follows from Theorem 2(i) that $\Sigma\,\theta^r(P_\nu)$ is a rational function of c for each r, so that the same is true for the elementary symmetric functions of the $\theta(P_\nu)$. Thus θ satisfies an equation of degree n whose coefficients are rational functions of ψ; and hence the minimal equation connecting θ and ψ has degree at most n in θ and hence by symmetry has degree at most m in ψ.

Now suppose that S is connected and that $F(\theta, \psi) = 0$ is the minimal equation. If F is not irreducible then we can write $F = F_1 F_2$ for some polynomials F_1 and F_2. Choose a point P on S and let N be the canonical neighbourhood of P associated with some local variable at P; then

$$F_1(\theta, \psi)F_2(\theta, \psi) = 0 \text{ in } N,$$

and so (by the analogous result for the complex plane) either $F_1(\theta, \psi) = 0$ in N or $F_2(\theta, \psi) = 0$ in N. Assume the former; then by analytic continuation $F_1(\theta, \psi) = 0$ everywhere on S, which contradicts the minimality of F. This completes the proof of the Lemma.

Theorem 4. The meromorphic functions on a compact connected Riemann surface S form a finitely generated field of transcendence degree 1 over **C**. Given any two points P_1 and P_2 on S, there is a meromorphic function ψ on S such that $\psi(P_1) \neq \psi(P_2)$. Moreover, given any point P on S there is a meromorphic function ϕ on S which is a local variable at P.

The difficult part of the proof is the second sentence, that there are enough meromorphic functions on S; in many practical cases this is irrelevant, since if one is presented with an explicit S it is likely to come equipped with plenty of functions. Given a non-constant meromorphic ψ on S, the proof of the first sentence is easy. For suppose that ψ has valence $n > 0$; then by Lemma 3 any meromorphic function on S is algebraic of degree at most n over $\mathbf{C}(\psi)$. Since the characteristic is zero, it follows from standard results in field theory that the field of all meromorphic functions on S is algebraic of degree at most n over $\mathbf{C}(\psi)$.

Corollary. Any compact connected Riemann surface can be regarded as an irreducible non-singular algebraic curve over **C**, and vice versa.

Because the functions on S separate points, S can be reconstructed from a knowledge of the field of meromorphic functions on it; and this field is of the sort that corresponds to an irreducible curve. The curve is non-singular because the field contains a local variable at each point. For the converse, we need only check that the Riemann surface corresponding to an irreducible curve is connected; if this were not so, the connected components of the Riemann surface would give rise to components of the curve.

The additive group of divisors on S is the free abelian group whose generators are the points of S. Let ψ be a meromorphic function on S which is not identically zero on any connected component of S, and let $P_1, \ldots, P_n$ be the poles and $Q_1, \ldots, Q_n$ be the zeros of ψ, repeated according to their multiplicities; then the divisor of ψ is defined to be

$$(\psi) = Q_1 + \ldots + Q_n - P_1 - \ldots - P_n.$$

The divisor of a differential ω is defined in a similar way and is written as (ω), though of course a differential need not have as many poles as zeros. Taking divisors is a homomorphism from the multiplicative group of non-zero meromorphic functions on S to the additive group of divisors. The kernel of this homomorphism is just the functions with no poles or zeros; these have valence 0 and are therefore constant on each connected component of S. In particular, if S is connected the divisor of a function on S determines the function up to multiplication by a non-zero constant. The image of the homomorphism, that is the group of divisors of functions, is called the group of principal divisors; and two divisors are said to be linearly equivalent if their difference is a principal divisor. The degree of a divisor $\Sigma n_\nu P_\nu$ is defined to be Σn_ν; thus linearly equivalent divisors have the same degree, but not necessarily vice versa. A divisor $\Sigma n_\nu P_\nu$ is said to be positive if every $n_\nu \geq 0$; this induces on the group of divisors a partial ordering which is compatible with addition. Since the quotient of two differentials is a function, and the product of a function and a differential is a differential, it is easy to see that the divisors of differentials precisely fill a linear equivalence class; this is called the canonical class and every divisor in it is called a canonical divisor.

There are two key theorems which tell one something about the structure of the group of divisors, the Riemann-Roch theorem which gives information about principal divisors and Abel's theorem which describes the group of divisor classes modulo linear equivalence. To state the Riemann-Roch theorem we need some notation. Let S be a connected compact Riemann surface, and $\mathfrak{a}$ a divisor on S. Denote by $L(\mathfrak{a})$ the **C**-vector space consisting of those meromorphic functions ψ on S for which $(\psi) + \mathfrak{a} \geq 0$, together with the zero function, and write $l(\mathfrak{a}) = \dim L(\mathfrak{a})$; thus $l(\mathfrak{a})$ depends only on the linear equivalence class of $\mathfrak{a}$. Since $\deg((\psi)) = 0$ for any non-zero function ψ on S, $L(\mathfrak{a}) = \{0\}$ and $l(\mathfrak{a}) = 0$ whenever $\deg(\mathfrak{a}) < 0$.

Theorem 5 (Riemann-Roch). Let S be a compact connected Riemann surface. Then there is an integer $g \geq 0$, depending only on S, such that

$$l(\mathfrak{a}) = \deg(\mathfrak{a}) + 1 - g + l(\mathfrak{k} - \mathfrak{a}) \tag{1}$$

for any divisor $\mathfrak{a}$ and any canonical divisor $\mathfrak{k}$.

The condition that S should be connected may be dispensed with, if we suitably modify the definition of $L(\mathfrak{a})$ and allow g to be negative. However this is not a real generalization, for the equation (1) for general S can be obtained by addition of the corresponding equations for the connected components of S.

Corollary 1. $\operatorname{Deg}(\mathfrak{k}) = 2g - 2$ and $l(\mathfrak{k}) = g$, if S is connected.

Proof. Applying Theorem 5 to $\mathfrak{k} - \mathfrak{a}$ instead of $\mathfrak{a}$ gives

$$l(\mathfrak{k} - \mathfrak{a}) = \deg(\mathfrak{k}) - \deg(\mathfrak{a}) + 1 - g + l(\mathfrak{a}),$$

and comparing this with (1) gives $\deg(\mathfrak{k}) = 2g - 2$. Now take $\mathfrak{a} = \mathfrak{k}$ and note that $l(0) = 1$ because $L(0) = \mathbf{C}$; thus $l(\mathfrak{k}) = g$.

The Riemann-Roch theorem is a duality theorem, relating $l(\mathfrak{a})$ and $l(\mathfrak{k} - \mathfrak{a})$; it can be written in self-dual form but nothing is gained by doing so. In the special case when $\deg(\mathfrak{a}) > 2g - 2$ we have $\deg(\mathfrak{k} - \mathfrak{a}) < 0$ and hence $l(\mathfrak{k} - \mathfrak{a}) = 0$; so (1) takes the form

$$l(\mathfrak{a}) = \deg(\mathfrak{a}) + 1 - g \quad \text{if} \quad \deg(\mathfrak{a}) > 2g - 2, \tag{2}$$

which is the form in which it is most frequently used.

Corollary 2. The differentials of the first kind form a **C**-vector space of dimension g, provided S is connected.

Proof. This is just the equation $l(\mathfrak{k}) = g$ in a new form. For let ω be a given non-zero differential; then the differentials are just the $\psi\omega$ where ψ runs through all meromorphic functions on S, and $\psi\omega$ is a differential of the first kind if and only if ψ is in $L((\omega))$, which has dimension g.

The statements of both these Corollaries need to be modified if S is not connected; in that case both $l(\mathfrak{k})$ and the dimension of the space of differentials of the first kind are equal to $(g - 1)$ plus the number of connected components of S.

The integer g in Theorem 5 is called the genus of S; if S is connected its genus is equal to the g which we have already defined topologically by the condition that $H_1(S, \mathbf{Z})$ is a free abelian group on 2g generators. To prove this we consider maps from one Riemann surface to another. Let S_1 and S_2 be compact connected Riemann surfaces; then a map $f : S_1 \to S_2$ is said to be holomorphic if for any point P_1 on S_1 and any local variable ϕ_2 at $f(P_1)$ on S_2 the function $\phi_2 \circ f$ is holomorphic at P_1. The point P_1 is said to be a ramification point of order r for f if $\{\phi_2 \circ f - \phi_2 \circ f(P_1)\}$ has a zero of order r at P_1 and $r > 1$; it follows from the compactness of S_1 that if f is non-constant it has only finitely many points of ramification. Moreover if P_2 is a point of S_2 then the degree of $f^{-1}(P_2)$ does not depend on the choice of P_2 provided that points of ramification in $f^{-1}(P_2)$ are taken with multiplicities equal to their orders of ramification; if this degree is n then we say that f is an n-to-1 map.

Theorem 6. _Let_ $f : S_1 \to S_2$ _be a non-constant holomorphic n-to-1 map between two compact connected Riemann surfaces, and let_ $r_1, \ldots, r_m$ _be the orders of the ramification points of_ f. _Then_

$$2g_1 - 2 = n(2g_2 - 2) + \Sigma(r_\mu - 1) \tag{3}$$

where g_1 _is the genus of_ S_1 _and_ g_2 _the genus of_ S_2.

Proof. Let ω_2 be a differential on S_2 and assume for convenience of description that none of the images of points of ramification of f are zeros or poles of ω_2. By means of f we can pull ω_2 back to a differential ω_1 on S_1. With the notation above, if P_1 is a point of S_1 which is not a ramification point of f then $\phi_2 \circ f$ is a local variable at P_1; and it follows that ω_1 has a zero (pole) at P_1 if and only if ω_2 has a zero (pole) at $f(P_1)$, and both zeros (poles) have the same multiplicity. On the other hand, if P_1 is a ramification point of order r then $d(\phi_2 \circ f)$ has a zero of order $r - 1$ at P_1 and hence so has ω_1. Thus the divisor of ω_1 is equal to the sum of the pull-back of the divisor of ω_2 and all the ramification points of f, a point of order r being taken with multiplicity $r - 1$. To prove (3) we now evaluate the degree of (ω_1) both directly and through this decomposition.

In particular let S_2 be the complex plane including the point at infinity; then f can be any non-constant meromorphic function on S_1 and n is the valence of f. Moreover $g_2 = 0$ since the differential dz has a double pole at infinity; so (3) reduces to

$$2g_1 = 2 - 2n + \Sigma(r_\mu - 1). \tag{4}$$

If one now constructs a triangulation of S_2 which has all the images of ramification points among its vertices, and lifts it to a triangulation of S_1, a straightforward calculation shows that the rank of $H_1(S_1, \mathbf{Z})$ is $2g_1$.

In the classification of connected Riemann surfaces, the genus is the only discrete-valued invariant that occurs; all connected Riemann surfaces of the same genus form a single continuous system. How can one most easily calculate the genus of a given connected Riemann surface? If the surface is defined topologically then one should use the topological description of the genus; in particular, if the surface is given with a triangulation, as in modular function theory for example, one can use the formula

$$2g - 2 = e - v - f$$

where the triangulation has v vertices, e edges and f faces. If the surface is defined by means of its function field one should use either $\deg((\omega)) = 2g - 2$ for some suitably chosen differential ω or else one of the covering formulae (3) and (4); the various formulae that occur in algebraic geometry can be deduced without difficulty from these.

Now let Ω denote the **C**-vector space of differentials of the first kind on a compact connected Riemann surface S of genus $g > 0$. By Theorem 1 the pairing

$$(\omega, \Gamma) \to \int_\Gamma \omega$$

induces a map

$$\Omega \times H_1(S, \mathbf{Z}) \to \mathbf{C}$$

which is **C**-linear in the first argument and additive in the second; so this map induces a canonical homomorphism

$$H_1(S, \mathbf{Z}) \to \Omega^* = \mathrm{Hom}(\Omega, \mathbf{C}). \qquad (5)$$

Call the image of this map Λ; it is obviously a free abelian group on at most 2g generators. Now let O be a fixed point and P an arbitrary point of S; to an arc OP corresponds an element of Ω^* given by $\omega \to \int_O^P \omega$. If we change the arc OP, leaving O and P the same, this element of Ω^* is changed by an element of Λ; thus to P corresponds an element of Ω^*/Λ, and by additivity this correspondence can be extended to a homomorphism

$$\{\text{Divisors on } S\} \to \Omega^*/\Lambda. \qquad (6)$$

If we restrict this homomorphism to divisors of degree zero, it no longer depends on the choice of the point O; and it is just what we need to pick out the principal divisors.

Theorem 7 (Abel). *With the notation above,* Λ *is a lattice in* Ω^* *- that is, a free abelian group on* $2g$ *generators which spans* Ω^* *considered as a real vector space. Moreover the homomorphism (6), restricted to divisors of degree zero, is onto and its kernel is just the group of principal divisors.*

Another way of stating the first sentence is to say that (5) induces a homeomorphism between $H_1(S, Z)$ with the discrete topology and its image Λ with the topology induced on it as a subset of Ω^*.

Abel's theorem identifies the group of divisor classes of degree zero with the complex torus Ω^*/Λ of dimension g, which is also called the Jacobian of S. Not surprisingly, this complex torus is actually an abelian manifold - that is, it has enough meromorphic functions on it to separate points. This will be proved in §9; for the proof we shall need further information about Λ or, which comes to the same thing, about the periods associated with S. This information is contained in *Riemann's relations*; to state them it is convenient to choose a normalized base for $H_1(S, Z)$. So let $\Gamma_1, \ldots, \Gamma_g, \Gamma'_1, \ldots, \Gamma'_g$ be closed curves on S such that

(i) the corresponding classes generate $H_1(S, Z)$; and

(ii) no two of these curves have a point in common,

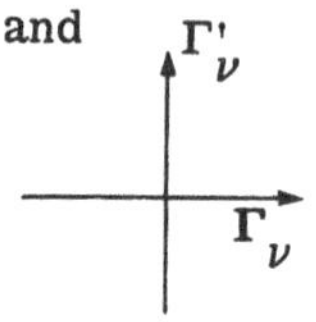

except that for each ν the curves Γ_ν and Γ'_ν have one common point at which they cross with the orientations shown in the diagram.

Such curves can be found, for if we consider S as a sphere with g handles we can choose Γ_ν to go once round the νth handle and Γ'_ν to run along the νth handle and back along the surface of the sphere. By considering intersection numbers we see first that the homology classes of these $2g$ curves are linearly independent and then that they form a base for $H_1(S, Z)$. Moreover any closed curve which does not meet any Γ_ν or Γ'_ν must be homologically trivial.

Theorem 8. (i) *Let* ω_1, ω_2 *be differentials of the first kind on* S *and let* $c_{\mu\nu}, c'_{\mu\nu}$ *be the periods of* ω_μ *with respect to* Γ_ν, Γ'_ν

respectively. Then

$$\Sigma(c_{1\nu}c'_{2\nu} - c'_{1\nu}c_{2\nu}) = 0. \tag{7}$$

(ii) Let ω be a non-zero differential of the first kind on S and let c_ν, c'_ν be its periods with respect to Γ_ν, Γ'_ν respectively. Then

$$\operatorname{Im} \Sigma \bar{c}_\nu c'_\nu > 0. \tag{8}$$

Proof. (i) Cut the Riemann surface S along all the Γ_ν and Γ'_ν and call the result S^*. The boundary of S^* consists of g pieces, a typical one of which consists of Γ_ν and Γ'_ν each taken twice, as in the figure; as one goes round this piece of the boundary one traverses each of Γ_ν and Γ'_ν once in each direction. The integral of ω_1 round any closed curve in S^* is zero, since such a curve is homologically trivial in S; so ω_1 has an indefinite integral f_1 on S^* which is a single-valued holomorphic function.

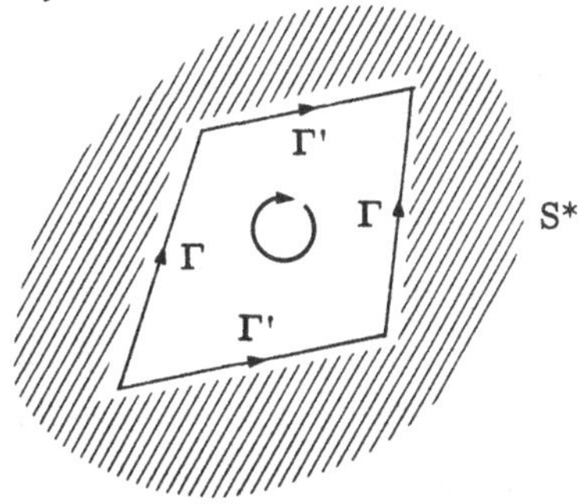

Moreover if ω^* is any holomorphic differential on S^* then the integral of ω^* round the boundary of S^*, each piece of the boundary being traversed so as to leave S^* on the left, is zero by Cauchy's theorem. In particular this holds for $\omega^* = f_1\omega_2$. But to any point of Γ_ν correspond two points of the boundary of S^*, and the values of f_1 at these two points differ by $c'_{1\nu}$ because any path in S^* that connects them is homologous in S to Γ'_ν. Taking account of directions, the integral of $f_1\omega_2$ round the two parts of the boundary of S^* which correspond to Γ_ν is

$$-c'_{1\nu}\int_{\Gamma_\nu}\omega_2 = -c'_{1\nu}c_{2\nu};$$

and similarly the integral round the two parts that correspond to Γ'_ν is $c_{1\nu}c'_{2\nu}$. Summation over the g pieces of the boundary now gives (7).

(ii) Define S^* as before and let f be an indefinite integral of ω on S^*. By a calculation like that of (i), the integral of $\bar{f}\omega/2i$ round the boundary of S^* is equal to

$$\Sigma(\bar{c}_\nu c'_\nu - c_\nu \bar{c}'_\nu)/2i = \mathrm{Im}\, \Sigma \bar{c}_\nu c'_\nu ;$$

so we need only prove that the integral is strictly positive. Write $f = u + iv$; then

$$\bar{f}\omega/2i = d(u^2 + v^2)/4i + \tfrac{1}{2}(udv - vdu).$$

By Green's theorem we obtain

$$\int \bar{f}\omega/2i = \tfrac{1}{2}\int(udv - vdu) = \int\int dudv$$

where the simple integrals are taken over the boundary of S^* and the double integral is over S^*. The double integral is positive since the integrand at any point P is equal to $|df/d\phi|^2 d\xi d\eta$ where $\phi = \xi + i\eta$ is a local variable at P. This proves (8).

By means of the Riemann-Roch theorem we can put into canonical form the function field of a Riemann surface of small genus. Consider first the case $g = 0$. Let P be any point of S; then $l(P) = 2$ by (2) and hence there is a non-constant meromorphic function z on S whose only singularity is a simple pole at P. By Lemma 3 the field of meromorphic functions on S is just $\mathbf{C}(z)$, and S can be identified with the z-plane including the point at infinity. We have already seen in the proof of (4) that the z-plane has genus 0; what we have now shown is that it is essentially the only connected compact Riemann surface of genus 0.

Consider now the case $g = 1$ and again let P be any point of S. Now (2) implies $l(P) = 1$, $l(2P) = 2$ and $l(3P) = 3$; so there exist meromorphic functions u, v on S with no singularity other than P and with respectively a double and a triple pole at P. The seven functions u^3, v^2, uv, u^2, v, u and 1 lie in L(6P) which has dimension 6; so they must be linearly dependent over $\mathbf{C}$. Moreover the linear dependence relation must involve both u^3 and v^2, for otherwise one of its terms would have a pole of strictly greater order than any of the others. By replacing v by $v + au + b$ for suitable a, b we can get rid of the terms in uv and v; and then by replacing u by $u + c$ we can get rid of the term in u^2. Multiplying u or v by a suitable constant now reduces the linear dependence relation to the form

$$v^2 = 4u^3 - g_2 u - g_3 \qquad (9)$$

for some constants g_2 and g_3. (The notation, and the factor 4, are chosen for historical reasons which will be explained in §2.) The only freedom left to us is to replace u and v by $c^2 u$ and $c^3 v$ respectively, for some $c \neq 0$; this would replace g_2 by $c^{-4} g_2$ and g_3 by $c^{-6} g_3$.

The function field of S is at most quadratic over $\mathbf{C}(u)$, by the argument of Theorem 4; so it is precisely $\mathbf{C}(u, v)$. Moreover the right hand side of (9) splits into three distinct linear factors, for if it were of the form $4(u - \alpha)^2(u - \beta)$ we could write $w = v/2(u - \alpha)$ whence $u = \beta + w^2$ and w would have a simple pole at P and no other pole, contrary to $l(P) = 1$. The condition that the right hand side of (9) splits into three distinct factors is $g_2^3 - 27g_3^2 \neq 0$; we therefore write

$$j = 1728 g_2^3/(g_2^3 - 27g_3^2)$$

so that j is finite and depends only on S and perhaps on the choice of P. (We shall see below that in fact it does not depend on the choice of P.) Conversely, consider the function field $\mathbf{C}(u, v)$ where u, v are connected by (9) with $g_2^3 - 27g_3^2 \neq 0$. It can be verified that du/v is a differential which has no poles or zeros; indeed this is implicitly proved in the next paragraph. Hence $g = 1$ by Corollary 1 of Theorem 5. Moreover du/v is the unique differential of the first kind, up to multiplication by a constant. By Theorem 5, to any divisor $\mathfrak{a}$ of degree 0 on S there is a function f, unique up to multiplication by a constant, such that $(f) + \mathfrak{a} + P \geq 0$; and hence there is a unique point Q on S such that $(f) + \mathfrak{a} + P = Q$. This is the same as saying that $\mathfrak{a}$ is linearly equivalent to $Q - P$; and in this way we identify S with its Jacobian, the identification $S \approx \mathbf{C}/\Lambda$ being induced by

$$Q \to \int_P^Q du/v$$

which is a holomorphic map. On the one hand this means that S has a natural structure of abelian group once one has chosen a base point P; on the other hand it means that a change of base point merely corresponds to a translation on $\mathbf{C}/\Lambda$. In particular, j does not depend on the choice of P. Thus any Riemann surface of genus 1 corresponds to an equation of the form (9), and such surfaces are classified by the points of the affine line.

Now consider those Riemann surfaces of genus $g > 0$ which admit a meromorphic function u of valence 2. The function field of such a surface cannot be $\mathbf{C}(u)$ so it must be a quadratic extension of $\mathbf{C}(u)$. We can therefore suppose it to be $\mathbf{C}(u, v)$ where $v^2 = f(u)$ for some polynomial $f(u) = u^n + \ldots$ of degree n with no repeated root. Now u is a local variable at any point of S except for the n points at which $v = 0$, where v is a local variable, and the one or two points at infinity. If n is odd there is one point at infinity and a local variable there is $u^{(n-1)/2}v$; if n is even there are two points at infinity and a local variable at each of them is $1/u$. It is now easy to check that the differential du/v has no poles or zeros except perhaps at infinity; it has a zero of order $n - 3$ at infinity if n is odd, and two zeros each of order $\frac{1}{2}n - 2$ if n is even. So the genus of S is $\frac{1}{2}(n - 1)$ if n is odd and $\frac{1}{2}(n - 2)$ if n is even. For $g > 1$ such a surface is called <u>hyperelliptic.</u>

Every Riemann surface of genus 2 is hyperelliptic. Indeed, let $\mathfrak{k}$ be a canonical divisor; then $\deg(\mathfrak{k}) = 2$ and $l(\mathfrak{k}) = 2$, so that $\mathfrak{k}$ is linearly equivalent to a positive divisor. Without loss of generality we may therefore assume $\mathfrak{k}$ positive. There is a non-constant function u on S with $(u) + \mathfrak{k} \geq 0$. This u has at most two poles and therefore has valence 2; valence 1 is impossible since a surface with a function of valence 1 must have genus 0. Using the results of the last paragraph, we see that the function field of S has the form $\mathbf{C}(u, v)$ with $v^2 = f(u)$, where f is a polynomial of degree 5 or 6 with no repeated root.

For any $g > 2$ there exist connected Riemann surfaces of genus g which are not hyperelliptic; but I do not know an easy proof of this fact.

2. Doubly periodic functions

In this section we consider the complex torus $\mathbf{C}/\Lambda$ of complex dimension 1, where Λ is a lattice in $\mathbf{C}$; this is a Riemann surface of genus 1, but we shall develop its theory directly rather than quote the results of §1. Let λ_1, λ_2 be a pair of generators of Λ, ordered so that $\mathrm{Im}(\lambda_1/\lambda_2) > 0$, and let z_0 be a point of $\mathbf{C}$; denote by Π the interior of the parallelogram whose vertices are z_0, $z_0 + \lambda_2$, $z_0 + \lambda_1 + \lambda_2$, $z_0 + \lambda_1$ and by $\boldsymbol{\Gamma}$ the boundary of Π taken anticlockwise, as in the

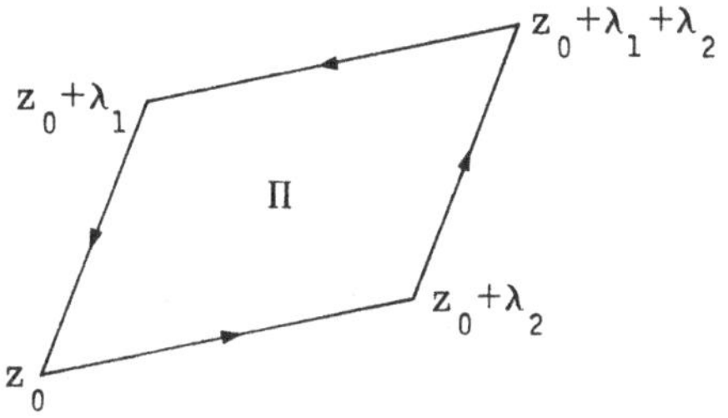

figure. In what follows we shall always suppose z_0 so chosen that none of the functions we are concerned with has a pole or zero on Γ; apart from this, z_0 is of no importance. By a doubly periodic function or elliptic function with respect to Λ we shall mean a meromorphic function $f(z)$ on $\mathbf{C}$ such that

$$f(z + \lambda) = f(z) \quad \text{for all } \lambda \text{ in } \Lambda;$$

this is the same as saying $f(z + \lambda_1) = f(z + \lambda_2) = f(z)$. Elliptic functions are so called because they were first obtained by inverting the integrals involved in finding the length of an arc of an ellipse.

Theorem 9. Let $f(z)$ be a doubly periodic function not identically zero, and let $P_1, \ldots, P_m$ be its poles and $Q_1, \ldots, Q_n$ its zeros in Π, account being taken of multiplicity. Then $m = n$ and

$$P_1 + \ldots + P_m \equiv Q_1 + \ldots + Q_n \bmod \Lambda.$$

Proof. We have

$$\int_\Gamma \frac{f'(z)}{f(z)}\, dz = 2\pi i(n - m), \qquad \int_\Gamma \frac{zf'(z)}{f(z)}\, dz = 2\pi i(\Sigma Q_\nu - \Sigma P_\mu). \quad (10)$$

In each integral we compare the contributions of opposite sides of the parallelogram. Since $f'(z)/f(z)$ is doubly periodic, we have

$$\int_{z_0}^{z_0+\lambda_2} \frac{f'(z)}{f(z)}\, dz + \int_{z_0+\lambda_1+\lambda_2}^{z_0+\lambda_1} \frac{f'(z)}{f(z)}\, dz$$

$$= \int_{z_0}^{z_0+\lambda_2} \left\{\frac{f'(z)}{f(z)} - \frac{f'(z+\lambda_1)}{f(z+\lambda_1)}\right\} dz = 0;$$

adding the similar equation for the other two sides and comparing with the first equation (10), we obtain $m = n$. Similarly

$$\int_{z_0}^{z_0+\lambda_2} \frac{zf'(z)}{f(z)} dz + \int_{z_0+\lambda_1+\lambda_2}^{z_0+\lambda_1} \frac{zf'(z)}{f(z)} dz$$
$$= \int_{z_0}^{z_0+\lambda_2} \frac{-\lambda_1 f'(z)}{f(z)} dz = [-\lambda_1 \log f(z)]_{z_0}^{z_0+\lambda_2} = 2\pi i n_1 \lambda_1$$

for some integer n_1. Adding the similar equation for the other two sides and comparing with the second equation (10), we obtain

$$\Sigma Q_\nu - \Sigma P_\mu = n_1\lambda_1 + n_2\lambda_2 \equiv 0 \bmod \Lambda.$$

This completes the proof of the Theorem. It has a converse, which will be stated and proved as Theorem 10; the two together are just Abel's theorem for $\mathbf{C}/\Lambda$, since the differentials of the first kind on $\mathbf{C}/\Lambda$ correspond to the constant multiples of dz.

We cannot have $m = n = 1$ in Theorem 9, since the one pole and the one zero of $f(z)$ in Π would have to coincide; so the simplest non-constant doubly periodic function that we can hope to construct will have double poles at the points of Λ and no other singularities. Let $f(z)$ be such a function; then $f(z) - f(-z)$ is doubly periodic with at most simple poles at the points of Λ, and therefore must be constant - and since it is an odd function of z this means it must be zero. Thus $f(z)$ is even, and by considering $af + b$ instead of f we can normalize it so that $f(z) = z^{-2} + O(z^2)$ near the origin. This is in fact enough to show us how to construct $f(z)$; but there is a slight difficulty over convergence which leads us to construct its derivative first. This satisfies $f'(z) = -2z^{-3} + O(z)$; so write

$$\wp'(z) = \wp'(z; \Lambda) = \wp'(z; \lambda_1, \lambda_2) = -2\Sigma(z + \lambda)^{-3} \tag{11}$$

where the sum is over all λ in Λ. Here there are no problems about convergence; indeed for this and the sums that occur in (12) and (14) one can easily prove the following. Let U be any compact subset of $\mathbf{C}$ and delete from the sum the finitely many terms that have a pole in U; then the sum that is left is uniformly absolutely convergent in U. This result is enough to justify all the formal manipulations that follow.

Clearly $\wp'(z)$ is odd and doubly periodic. Its integral, which is single-valued because $\wp'(z)$ has residue 0 at every pole, is

$$\wp(z) = z^{-2} + \int_0^z \{\wp'(u) + 2u^{-3}\} du = z^{-2} + \Sigma' \{(z+\lambda)^{-2} - \lambda^{-2}\} \quad (12)$$

where the prime denotes that the sum is over all non-zero λ in Λ. Clearly $\wp(z)$ is even. One way to prove that $\wp(z)$ is doubly periodic is by rearranging the series, but it is tedious to justify this; instead we proceed as follows. The function $g(z) = \wp(z + \lambda_1) - \wp(z)$ is constant because its derivative vanishes; and $g(-\frac{1}{2}\lambda_1) = 0$ because $\wp(z)$ is even. So $\wp(z + \lambda_1) = \wp(z)$, and similarly $\wp(z + \lambda_2) = \wp(z)$. This function $\wp(z)$, which is the simplest possible doubly periodic function, is called the <u>Weierstrass</u> $\wp$-<u>function.</u>

Now $\wp'^2 - 4\wp^3$ is an even doubly periodic function which has at most poles of order 2 at the points of Λ and no other singularities. Subtracting a suitable multiple of $\wp$ we obtain a holomorphic doubly periodic function f(z) say; and f(z) - f(0) must vanish identically by Theorem 9. So there is an equation of the form

$$\wp'^2 = 4\wp^3 - g_2\wp - g_3 \quad (13)$$

where g_2 and g_3 depend only on Λ. Moreover the function on $\mathbf{C}/\Lambda$ corresponding to $\wp(z)$ has valence 2, so that as in the proof of Theorem 4 the field of doubly periodic functions is at most a quadratic extension of $\mathbf{C}(\wp)$. Hence this field is precisely $\mathbf{C}(\wp, \wp')$ with $\wp$ and $\wp'$ algebraically related by (13).

Since $\wp(z)$ has residue 0 at every pole, it has a single-valued integral which we write after a change of sign as the Weierstrass zeta-function

$$\zeta(z) = z^{-1} - \int_0^z \{\wp(u) - u^{-2}\} du = z^{-1} + \Sigma' \{(z+\lambda)^{-1} - \lambda^{-1} + z\lambda^{-2}\}; \quad (14)$$

here as in (12) the prime denotes that the sum is over all non-zero λ in Λ. The function $\zeta(z + \lambda) - \zeta(z)$ is constant for any fixed λ because its derivative vanishes; but there is no longer any reason why this constant should vanish. Define η_1, η_2 by

$$\zeta(z + \lambda_1) - \zeta(z) = \eta_1, \quad \zeta(z + \lambda_2) - \zeta(z) = \eta_2; \tag{15}$$

a trivial induction argument shows that

$$\zeta(z + n_1\lambda_1 + n_2\lambda_2) - \zeta(z) = n_1\eta_1 + n_2\eta_2$$

for any integers n_1 and n_2. Moreover an argument similar to that in the proof of Theorem 9 shows that

$$\int_\Gamma \zeta(z)dz = 2\pi i = \lambda_1\eta_2 - \lambda_2\eta_1; \tag{16}$$

the right-hand equality here is known as Legendre's relation.

The singularities of $\zeta(z)$ are simple poles of residue 1 at each point of Λ; so the integral of $\zeta(z)$ will have a logarithmic branch point at each of these points. We obtain a single-valued function by taking the exponential of this integral; the result is the Weierstrass sigma-function

$$\sigma(z) = \exp\{\log z + \int_0^z(\zeta(u) - u^{-1})du\} = z\Pi'\{\frac{z+\lambda}{\lambda}\exp(-\frac{z}{\lambda} + \frac{z^2}{2\lambda^2})\}$$

where the product is taken over all non-zero λ in Λ. Clearly $\sigma(z)$ is an odd holomorphic function whose only zeros are simple ones at the points of Λ. Integrating (15) and exponentiating gives

$$\sigma(z + \lambda_\nu) = B_\nu\sigma(z)\exp(z\eta_\nu) \tag{17}$$

for $\nu = 1, 2$ and some constants B_ν. Setting $z = -\frac{1}{2}\lambda_\nu$ and remembering that $\sigma(z)$ is odd and does not vanish at $\frac{1}{2}\lambda_\nu$, we obtain the value of B_ν; substituting this back into (17) we obtain

$$\sigma(z + \lambda_\nu) = -\sigma(z)\exp\{\eta_\nu(z + \tfrac{1}{2}\lambda_\nu)\} \tag{18}$$

for $\nu = 1, 2$. From this we can obtain the general functional equation

$$\sigma(z + \lambda) = (-1)^n\sigma(z)\exp\{\eta(z + \tfrac{1}{2}\lambda)\}$$

where $\lambda = n_1\lambda_1 + n_2\lambda_2$, $\eta = n_1\eta_1 + n_2\eta_2$, $n = n_1n_2+n_1+n_2$ and n_1, n_2

are integers. In many ways the sigma-function is the most important of the Weierstrass functions, the extra complications of its functional equation being counter-balanced by the simplicity of its divisor. This is illustrated by the proof that follows.

Theorem 10. For some $n > 1$ let $P_1, \ldots, P_n$ and $Q_1, \ldots, Q_n$ be points of Π, not necessarily all distinct but such that no P_μ is a Q_ν. Suppose also that

$$P_1 + \ldots + P_n \equiv Q_1 + \ldots + Q_n \bmod \Lambda.$$

Then there is a doubly periodic function whose poles in Π are just the P_ν and whose zeros in Π are just the Q_ν.

Proof. Choose points $R_1, \ldots, R_n$ such that $R_\nu \equiv P_\nu \bmod \Lambda$ and $\Sigma R_\nu = \Sigma Q_\nu$; then the function

$$f(z) = \frac{\sigma(z-Q_1) \ldots \sigma(z-Q_n)}{\sigma(z-R_1) \ldots \sigma(z-R_n)} \tag{19}$$

has the correct poles and zeros, and it is doubly periodic by (18).

An important generalization of the sigma-functions is given by theta-functions; these are the holomorphic functions $\theta(z)$ such that to any λ in Λ there corresponds an inhomogeneous linear function of z, say $F(z, \lambda)$, such that

$$\theta(z + \lambda) = \theta(z) \exp F(z, \lambda). \tag{20}$$

Clearly it is enough to check this for λ_1 and λ_2. Now assume that $\theta(z)$ is a theta-function not identically zero, and let $Q_1, \ldots, Q_n$ be its zeros in Π with the correct multiplicities; thus

$$\psi(z) = \theta(z) / \{\sigma(z - Q_1) \ldots \sigma(z - Q_n)\}$$

is a trivial theta-function, that is, a theta-function with no zeros.

Lemma 11. Any trivial theta-function has the form $\exp Q(z)$ where $Q(z)$ is a quadratic polynomial; conversely any such $\exp Q(z)$ is a trivial theta-function.

Proof. Let $\theta(z)$ be a trivial theta-function; then any branch of $\log \theta(z)$ is holomorphic in the z-plane and satisfies

$$\log \theta(z + \lambda_\nu) - \log \theta(z) = F(z, \lambda_\nu) = O(1 + |z|) \tag{21}$$

for $\nu = 1, 2$. Since $\log \theta(z)$ is bounded on Π and it takes only $O(1 + |z|)$ steps each $\pm\lambda_1$ or $\pm\lambda_2$ to go from any z to a point of Π, it follows from (21) that

$$\log \theta(z) = O(1 + |z|^2);$$

and this implies that $\log \theta(z)$ is a quadratic polynomial. The converse is trivial.

Lemma 11 and the argument above it shows that any theta-function can be simply expressed in terms of sigma-functions; but there are other ways of writing them down. First of all, we consider the number of zeros of $\theta(z)$ in Π - the zeros are periodic even though $\theta(z)$ itzelf is not. Let $F(z, \lambda_\nu) = a_\nu z + b_\nu$ for $\nu = 1, 2$; then

$$\theta'(z + \lambda_\nu)/\theta(z + \lambda_\nu) = \theta'(z)/\theta(z) + a_\nu$$

and so the number of zeros of $\theta(z)$ in Π is

$$\frac{1}{2\pi i} \int_\Gamma \frac{\theta'(z)}{\theta(z)} dz = \frac{\lambda_1 a_2 - \lambda_2 a_1}{2\pi i}$$

by an argument like that in the proof of Theorem 9. In particular

$$\lambda_1 a_2 - \lambda_2 a_1 = 2\pi i n \text{ for some integer } n \geq 0 \tag{22}$$

which is the number of zeros of $\theta(z)$ in Π.

We now show how to construct all theta-functions with given a_ν, b_ν satisfying (22). It is convenient to construct not $\theta(z)$ itself but

$$\psi(z) = \theta(z) \exp\{-\tfrac{1}{2} a_2 \lambda_2^{-1} z(z - \lambda_2) - b_2 \lambda_2^{-1} z\}$$

where the multiplier has been chosen to ensure

$$\psi(z + \lambda_2) = \psi(z). \tag{23}$$

The other functional equation for ψ has the form

$$\psi(z + \lambda_1) = \psi(z)\exp(az + b) \tag{24}$$

for some a, b; here $a = -2\pi in\lambda_2^{-1}$ by (22), and the value of b is not important. By (23) we can write

$$\psi(z) = \sum_{-\infty}^{\infty} c_\nu \exp(2\pi i\nu z/\lambda_2) \tag{25}$$

for some constants c_ν; and (24) now becomes

$$c_{\nu+n} = c_\nu \exp\{2\pi i\nu\lambda_1/\lambda_2 - b\}.$$

A simple induction argument shows that

$$c_{\nu+mn} = c_\nu \exp\{\pi im(m-1)n\lambda_1/\lambda_2 + 2\pi i\nu m\lambda_1/\lambda_2 - bm\}$$

for every integer m. Hence we can choose $c_0, \ldots, c_{n-1}$ arbitrarily (provided $n > 0$) and the other c_ν are uniquely determined. Since $\mathrm{Im}(\lambda_1/\lambda_2) > 0$ the c_ν die away rapidly as $|\nu| \to \infty$; in particular the series (25) is absolutely convergent.

It follows that for $n > 0$ the theta-functions with given a_ν, b_ν form an n-dimensional **C**-vector space. We could use this to give a proof of Theorem 10 which does not involve constructing the Weierstrass functions. Again, we can express $\sigma(z)$ as an infinite sum in this way, up to multiplication by a suitable constant. Unfortunately the constant involved can itself only be written as an infinite sum or product, and it is not possible to give a concise account of the calculations involved. Further information may be found in Weber [13] or Tannery and Molk [12].

3. Functions of several complex variables

Much of the theory of functions of several complex variables is a routine generalization of the corresponding definitions and results in the theory of functions of one complex variable. For example, let U be an open subset of $\mathbf{C}^n$; then a function $f(z_1, \ldots, z_n)$ is said to be <u>holomorphic</u> in U if it is differentiable at each point of U - that is, to any point $(\zeta_1, \ldots, \zeta_n)$ of U there correspond constants $B_1, \ldots, B_n$ such that

$$f(z_1, \ldots, z_n) = f(\zeta_1, \ldots, \zeta_n) + \Sigma B_\nu(z_\nu - \zeta_\nu) + o(\Sigma |z_\nu - \zeta_\nu|).$$

If we write $z_\nu = x_\nu + iy_\nu$, this is equivalent to f being a differentiable function of the $2n$ real variables x_ν, y_ν and satisfying the n Cauchy-Riemann equations

$$\partial f/\partial y_\nu = i\, \partial f/\partial x_\nu \qquad (\nu = 1, 2, \ldots, n).$$

It follows that f can be expanded about $(\zeta_1, \ldots, \zeta_n)$ as a multiple power series

$$f(z_1, \ldots, z_n) = \sum_0^\infty \ldots \sum_0^\infty a(m_1, \ldots, m_n)(z_1 - \zeta_1)^{m_1} \ldots (z_n - \zeta_n)^{m_n}$$

absolutely convergent in some neighbourhood of $(\zeta_1, \ldots, \zeta_n)$. Moreover the coefficients of this power series are given by a multiple Cauchy formula

$$a(m_1, \ldots, m_n) = \frac{1}{(2\pi i)^n} \int_{\Gamma_1} \ldots \int_{\Gamma_n} \frac{f(z_1, \ldots, z_n) dz_1 \ldots dz_n}{(z_1 - \zeta_1)^{m_1+1} \ldots (z_n - \zeta_n)^{m_n+1}}$$

where Γ_ν is a closed contour in the z_ν-plane surrounding ζ_ν and $U \supset D_1 \times \ldots \times D_n$ where D_ν is the closed disc in the z_ν-plane whose boundary is Γ_ν. However there is no analogue of a circle of convergence; indeed the domain of absolute convergence of a multiple power series can be quite complicated.

Let U be a connected open subset of $\mathbf{C}^n$. A function $g(z_1, \ldots, z_n)$ is said to be <u>meromorphic</u> on U if it is defined on a dense open subset of U and if for each point P of U there exist a neighbourhood $N(P)$ of P and two functions f_1 and f_2 holomorphic in $N(P)$ such that $g = f_1/f_2$ at every point of $N(P)$ at which g is defined. Two functions both meromorphic on U are identified if they are equal at every point at which they are both defined. By abuse of language one can speak of g having the value infinity at certain points, but even then (in contrast with the situation for meromorphic functions of one complex variable) g may have no definable value at certain points - consider for example $g = z_1/z_2$ at the origin.

In the theory of functions of one complex variable, each zero of a holomorphic function (and each zero or pole of a meromorphic function) is isolated and can be considered separately. Moreover, the functions which vanish at a given point P form a principal ideal in the ring of functions holomorphic at P, and that ring is a local ring. So one can speak of a function having a zero of given order at P, and that one integer gives the most important information about the behaviour of the function at P. With functions of $n > 1$ complex variables matters are more complicated, primarily because the set of zeros of a holomorphic function will have complex dimension $n - 1$, at least in an intuitive sense. Such a set does not naturally break up into local pieces; yet we cannot easily give any but local descriptions, and indeed the only easy description of the set of zeros of f will be by means of f itself. Intuitively one still thinks of the divisor of a meromorphic function as 'zeros minus poles', and a general divisor as an element of the free group generated by 'respectable' point sets of complex dimension $n - 1$; but formally it is necessary to proceed rather differently. For convenience we consider divisors on $\mathbf{C}^n$; the definitions can be transferred to complex manifolds of dimension n by means of local coordinates, much as in §1.

Let $\{U_\alpha\}$ be a covering of $\mathbf{C}^n$ by open sets, and for each α let f_α be a meromorphic function on U_α not identically zero. We say that the system $\{U_\alpha, f_\alpha\}$ is a description of a divisor if the following consistency condition holds:

If $U_\alpha \cap U_\beta$ is not empty then f_α/f_β and f_β/f_α are equal to holomorphic functions on $U_\alpha \cap U_\beta$.

In such a case, of course, the holomorphic functions will be inverses of each other and will therefore have no zeros in $U_\alpha \cap U_\beta$. So intuitively this means that f_α and f_β have the same poles and zeros in $U_\alpha \cap U_\beta$. To any refinement of the covering $\{U_\alpha\}$ there corresponds an obvious refinement of the description of a divisor. Since any two open coverings have a common refinement, in defining two descriptions of divisors as equivalent we can confine ourselves to the case when they are both based on the same covering. We shall say that the two descriptions $\{U_\alpha, f_\alpha\}$ and $\{U_\alpha, g_\alpha\}$ are <u>equivalent</u> if and only if f_α/g_α and g_α/f_α are equal to holomorphic functions on U_α for each α. An alternative definition, which works even if the two descriptions are not based on the same open covering, is to say that two descriptions are equivalent if their union is a description. Now a divisor on $\mathbf{C}^n$ is an equivalence class of descriptions of divisors. The divisor of a holomorphic or meromorphic function f on $\mathbf{C}^n$ is the equivalence class of the description $(\mathbf{C}^n, f)$. A divisor is said to be positive if it has a description $\{U_\alpha, f_\alpha\}$ such that f_α is holomorphic in U_α for each α; if so, every description of it has this property.

It is now a straightforward matter to develop the elementary theory of divisors and to show that they have all the properties which one expects by analogy with divisors on Riemann surfaces or in algebraic geometry - with two closely connected exceptions. It is not obvious how to prove that every divisor is the difference of two positive divisors, nor how to prove that two positive divisors have a highest common factor. For both these proofs, we need the Weierstrass Preparation Theorem, which tells us how far we can normalize a holomorphic function in a neighbourhood of a point P by multiplying it by a function which is holomorphic and non-zero at P. To simplify the notation, we take P to be the origin.

Theorem 12 (Weierstrass). Let $f(z_1, \ldots, z_n)$ be holomorphic in a neighbourhood of the origin, let $\phi(z_1)$ be the restriction of f to $z_2 = \ldots = z_n = 0$ and suppose that $\phi(z_1)$ does not vanish identically and has a zero of order $m \geq 0$ at $z_1 = 0$. Then there exist a function $g(z_1, \ldots, z_n)$ holomorphic and non-zero at the origin and a polynomial in z_1 of degree m,

$$p(z_1; z_2, \ldots, z_n) = z_1^m + z_1^{m-1} a_1(z_2, \ldots, z_n) + \ldots + a_m(z_2, \ldots, z_n)$$

where the $a_\mu(z_2, \ldots, z_n)$ are holomorphic at the origin, such that

$$f(z_1, \ldots, z_n) = g(z_1, \ldots, z_n) p(z_1; z_2, \ldots, z_n).$$

Moreover these conditions determine g and p uniquely for given f.

Proof. We can assume that $m > 0$ since if $m = 0$ the choice $p = 1$, $g = f$ is both forced and acceptable. Let Γ denote the circle $|z_1| = c$ in the z_1-plane, where $c > 0$ is chosen so that there are no zeros of $\phi(z_1)$ inside or on Γ except for the m-fold zero at the origin;

and let N be a connected open neighbourhood of the origin in the $(z_2, \ldots, z_n)$-space such that f has no zero on $\Gamma \times N$. For any $(z_2, \ldots, z_n)$ in N,

$$I(z_2, \ldots, z_n) = \frac{1}{2\pi i} \int_\Gamma f^{-1} \frac{\partial f}{\partial z_1} dz_1$$

is the number of zeros of f inside Γ, where f is considered as a function of z_1 alone. So this is equal to m at the origin and is integer-valued; on the other hand it is continuous in N. So it is equal to m for all $(z_2, \ldots, z_n)$ in N. Denote the m corresponding zeros of f by $\zeta_1, \ldots, \zeta_m$, these being complex numbers depending on $z_2, \ldots, z_n$. For any integer $r \geq 0$ we have

$$\zeta_1^r + \ldots + \zeta_m^r = \frac{1}{2\pi i} \int_\Gamma \frac{z_1^r}{f} \frac{\partial f}{\partial z_1} dz_1$$

and this is clearly a holomorphic function in N; hence so are the elementary symmetric functions of the ζ_μ. Now let $p(z_1)$ be the polynomial whose roots are the ζ_μ, and let D denote the interior of Γ in the z_1-plane. By writing it as an integral over Γ we see that f/p is holomorphic in $D \times N$, and by construction it is non-zero in $D \times N$; so p and $g = f/p$ have the properties stated in the Theorem. Moreover, any candidate for p must vanish at $\zeta_1, \ldots, \zeta_m$ as long as these lie in some suitable neighbourhood of the origin, and this is enough to prove uniqueness.

Corollary. Let $f(z_1, \ldots, z_n)$ be meromorphic in a neighbourhood of the origin. Then there is an open neighbourhood N of the origin and two functions u, v holomorphic in N and such that $f = u/v$ in N, with the following property. For any open subset U of N and any functions u_1, v_1 holomorphic in U and such that $f = u_1/v_1$ in U, there is a function ϕ holomorphic in U such that $u_1 = \phi u$ and $v_1 = \phi v$ in U.

Intuitively, this says that the fraction $u/v = f$ is in lowest possible terms everywhere in N. Of course, u and v are not uniquely determined.

Proof. We may assume that f does not vanish identically, for otherwise we can take $u = 0$ and $v = 1$. Let u_2 and v_2 be holomorphic functions in some neighbourhood of the origin such that $f = u_2/v_2$. Neither u_2 nor v_2 vanishes, so after a linear transformation on $z_1, \ldots, z_n$ if necessary, we can assume that the restrictions of u_2 and v_2 to $z_2 = \ldots = z_n = 0$ do not vanish. Applying Theorem 12 to both u_2 and v_2 we find that there is a neighbourhood $N = D \times N_1$ of the origin (where D is a disc in the z_1-plane and N_1 is an open set in the $(z_2, \ldots, z_n)$-space) and a decomposition $f = gu_3/v_3$ where g is holomorphic and non-zero in N and u_3, v_3 are holomorphic functions in N which are polynomials in z_1. Moreover, for $(z_2, \ldots, z_n)$ in N_1, the roots of u_3 and v_3 regarded as polynomials in z_1 all lie in D.

Let K be the field of functions of $z_2, \ldots, z_n$ meromorphic in N_1; then u_3 and v_3 regarded as elements of $K[z_1]$ have a highest common factor $p(z_1)$ and we can require $p(z_1)$ to have leading coefficient 1. For $(z_2, \ldots, z_n)$ in N_1 the roots of $p(z_1)$ all lie in D and so the coefficients of $p(z_1)$ are bounded; being also meromorphic they must be holomorphic in N_1. Thus we can divide u_3 and v_3 by p without altering the holomorphicity of their coefficients; in other words we can without loss of generality take u_3 and v_3 to be coprime as elements of $K[z_1]$. We now set $u = gu_3$ and $v = v_3$. Let $R(z_2, \ldots, z_n)$ be the resultant of u_3 and v_3; then provided $(z_2, \ldots, z_n)$ in N_1 is such that $R \neq 0$, the zeros of u_3 and v_3 are respectively the zeros and poles of f regarded as a function of z_1 in D. It follows that the decomposition $f = u/v$ does not depend on the choice of the original decomposition $f = u_2/v_2$; and $u_2/u = v_2/v$ is holomorphic in N.

Suppose we repeat the process after a linear transformation of coordinates, obtaining this time $f = u_4/v_4$; by the final remark in the last sentence, both $u/u_4 = v/v_4$ and $u_4/u = v_4/v$ are holomorphic at the origin, so the only effect is possibly to decrease N and to multiply both numerator and denominator by some invertible holomorphic function.

Now let $f = u_1/v_1$ in U and let P be any point of U. Since the decomposition $f = u/v$ is essentially independent of choice of coordinates, we may suppose the coordinates so chosen that u, v, u_1 and v_1 all satisfy the conditions of Theorem 12 at P. Now the polynomial parts of

u and v at P are coprime, since they are factors of u_3 and v_3 respectively; and so the canonical decomposition of f at P is obtained from u/v by multiplying numerator and denominator by a function holomorphic and non-zero at P. By the final remark in the second paragraph of this proof, $u_1/u = v_1/v$ is holomorphic at P; and since P is any point of U this completes the proof of the Corollary.

Theorem 13. Any divisor on a complex n-dimensional manifold can be written as the difference of two positive divisors.

Proof. Suppose that $\{U_\alpha, f_\alpha\}$ is a description of the divisor. To each point P of each U_α we can construct a neighbourhood $N_{\alpha P}$ of P contained in U_α and functions $u_{\alpha P}$ and $v_{\alpha P}$ holomorphic in $N_{\alpha P}$ such that $f_\alpha = u_{\alpha P}/v_{\alpha P}$ satisfies the conditions of the last Corollary in $N_{\alpha P}$. Clearly $\{N_{\alpha P}, f_\alpha\}$ is a description of our original divisor, so to prove the Theorem it is enough to show that $\{N_{\alpha P}, u_{\alpha P}\}$ and $\{N_{\alpha P}, v_{\alpha P}\}$ are descriptions of divisors. Suppose $N_{\alpha P}$ meets $N_{\beta Q}$, where α and β or P and Q may be the same. In the intersection we have

$$u_{\alpha P}/v_{\alpha P} = f_\alpha = f_\beta = u_{\beta Q}/v_{\beta Q}$$

and hence by the last Corollary $u_{\alpha P}/u_{\beta Q} = v_{\alpha P}/v_{\beta Q}$ is holomorphic and invertible in $N_{\alpha P} \cap N_{\beta Q}$. This is just the consistency condition we need.

A similar argument proves that two positive divisors have a highest common factor; but we shall not need this result in what follows.

It is convenient to include one more Lemma in this section. Even if we are initially given a covering with lots of good properties, these are likely to be lost during the processes described above; so it is desirable to know that we can refine any covering to a 'good' covering. Lemma 14 below is stated for a complex torus $\mathbf{C}^n/\Lambda$, where Λ is a lattice in $\mathbf{C}^n$, because this is the case which we shall need; there is an analogous result for $\mathbf{C}^n$ with 'finite' replaced by 'locally finite' which is proved in the same way. We shall say that a set U in $\mathbf{C}^n/\Lambda$ is an open hypersphere if its inverse image in $\mathbf{C}^n$ is the disjoint union of an open hypersphere

$$|z_1 - \zeta_1|^2 + \ldots + |z_n - \zeta_n|^2 < r^2$$

and its translations by elements of Λ. In other words, it is the image of an open hypersphere in $\mathbf{C}^n$ which is small enough for its translations not to overlap itself.

Lemma 14. Let $\{U_\alpha\}$ be a covering of the complex torus $\mathbf{C}^n/\Lambda$; then it can be refined to a finite covering by open hyperspheres.

Proof. To each point P of $\mathbf{C}^n/\Lambda$ we can choose a U_α containing P and an open hypersphere containing P which is contained in this U_α. The set of all these open hyperspheres forms a covering of $\mathbf{C}^n/\Lambda$ which refines the original covering; and because $\mathbf{C}^n/\Lambda$ is compact we can select from this a finite covering.

It is essential for the application that the new covering should be locally finite, that the open sets involved and all their intersections should be 'nice' sets, and that their inverse images should be disjoint unions of open balls. That the sets are open hyperspheres is just a device to simplify the construction of 'partitions of unity' in §4.

II · Theta functions and Riemann forms

This Chapter is mainly concerned with the question: when does a complex torus $T = \mathbf{C}^n/\Lambda$ have non-constant meromorphic functions on it, and how does one construct them when they exist? It is not possible to mimic the method used in §2 to construct the Weierstrass $\wp$-function, essentially because when $n > 1$ the set of poles of a multiply periodic meromorphic function on $\mathbf{C}^n$ does not break up naturally as a union of local objects; so a series expansion analogous to (11) or (12) can no longer be expected to converge, and indeed in general any neighbourhood of a given point will contain poles of infinitely many terms. Moreover, it is not likely that there will even be a description of the set of poles simpler than that given by the meromorphic function itself. We therefore fall back on the other method outlined in §2, by showing that any multiply periodic function can be written as the quotient of two theta-functions and by then constructing all theta-functions for Λ.

The first part of this program is carried through in §4. In the case $n = 1$ we could give a very simple proof of this result, because we had a good description of the possible divisors of multiply periodic functions and because we had a large enough supply of theta-functions; so from our stock of theta-functions we could construct a meromorphic theta-function with the same poles and zeros as our given doubly periodic function. This reduced the problem to the study of trivial theta-functions. For $n > 1$ this method does not work; instead we have to construct from the given multiply periodic function the theta-functions that we need. The ideas involved are really those of Hodge theory and of cohomology, but because one has so explicit a description of T the proof can be made to appear totally elementary; but this does make the motives for some of the steps obscure.

The second part of the program is complicated by the fact that Λ only admits non-trivial theta-functions if it satisfies a further condition;

this is the existence of a non-trivial Riemann form which is at least positive semi-definite. (This condition played no part in §2 because it is automatically satisfied when $n = 1$.) In §5 we prove that this condition is necessary; in §6 we assume that a Riemann form for Λ exists and construct all theta-functions associated with it. Both these sections contain other related results; in particular §6 contains a discussion of the function field of an abelian manifold, and the theorem of the projective embedding of an abelian manifold.

In Chapters II and III it will be convenient to denote by V the particular space $\mathbf{C}^n$ in which Λ is embedded, or sometimes the underlying real space $\mathbf{R}^{2n}$. The complex torus $V/\Lambda = \mathbf{C}^n/\Lambda$ will in general be denoted by T; but it will be denoted by A if it is known to be an abelian manifold - that is, if Λ admits a positive definite Riemann form.

4. Reduction to theta functions

In most of this section, we shall consider V as a real vector space spanned by the lattice Λ, and $T = V/\Lambda$ as a real torus; and unless otherwise stated functions on V or T will merely be complex-valued infinitely differentiable functions of the $2n$ real local variables x_ν and y_ν, where $z_\nu = x_\nu + iy_\nu$. What forces us to this is the fact that Lemma 16 below has no holomorphic analogue. It turns out also to be convenient to make the complex change of variables from x_ν, y_ν to z_ν, $\bar{z}_\nu$; the fact that the new variables are complex-valued does not invalidate any of the standard formalism. Moreover, a function f is holomorphic in an open domain U if and only if it satisfies the Cauchy-Riemann equations in U; and in terms of the new variables these take the convenient form

$$\partial f/\partial \bar{z}_\nu = 0 \quad \text{for } \nu = 1, 2, \ldots, n.$$

In view of this, one says that a differential r-form is holomorphic if it can be written

$$\sum_{i_1, \ldots, i_r} \cdots \sum f_{i_1, \ldots, i_r} dz_{i_1} \wedge \ldots \wedge dz_{i_r}$$

with all the $f_{i_1, \ldots, i_r}$ holomorphic. Again, an (r+s)-form is said to be of <u>type</u> (r, s) if it can be written

$$\sum_{i_1,\ldots,i_r,j_1,\ldots,j_s} \cdots \sum f_{i_1,\ldots,i_r,j_1,\ldots,j_s} dz_{i_1} \wedge \ldots \wedge dz_{i_r} \wedge d\bar{z}_{j_1} \wedge \ldots \wedge d\bar{z}_{j_s}$$

where now there is no restriction on the functions involved. If ω is of type (r, s) then $d\omega$ is the sum of a form of type (r+1, s) and a form of type (r, s+1). Moreover an r-form ω is holomorphic if and only if it is of type (r, 0) and $d\omega$ is of type (r+1, 0).

If Λ is a lattice in $\mathbf{C}^n$ then the definition of a theta-function with respect to Λ is the obvious generalization of that given in §2 for the case $n = 1$. It is a function $\theta(z)$ holomorphic in $\mathbf{C}^n$ and such that to each λ in Λ there corresponds a form $F(z, \lambda)$ inhomogeneous $\mathbf{C}$-linear in z and satisfying

$$\theta(z + \lambda) = \theta(z)\exp F(z, \lambda).$$

The divisor of θ is positive and periodic; θ is called a <u>trivial theta-function</u> if its divisor is null - that is, if $\theta(z)$ is never zero. In this case $\log \theta(z)$ is holomorphic, and an argument like that in the proof of Lemma 11 shows that the trivial theta-functions are just the $\exp Q(z)$ where $Q(z)$ is inhomogeneous quadratic in $z_1, \ldots, z_n$.

Lemma 15 (Poincaré). <u>Let</u> U <u>be a convex open subset of</u> $\mathbf{C}^n$ <u>and</u> ω <u>a 1-form on</u> U <u>such that</u> $d\omega = 0$. <u>Then there is a function</u> f <u>on</u> U <u>such that</u> $df = \omega$.

Proof. For some fixed point O in U write $f(P) = \int_O^P \omega$ where the integral is taken along any polygonal arc lying entirely in U; then to prove the Lemma we have only to show that f is well-defined - that is, independent of the path of integration. For this it is enough to show that $\int \omega = 0$ when the integral is taken over any closed polygon, and an obvious dissection reduces this to the case of the boundary of a triangle Δ. But by Stokes' theorem

$$\int_{\partial\Delta} \omega = \iint_\Delta d\omega = 0.$$

The result holds if U is any open ball, but the proof in this more general case involves topological complications.

Lemma 16. Let $\{U_1, \ldots, U_N\}$ be a covering of T by open hyperspheres and let $r \geq 0$ be an integer. Suppose that for each pair of subscripts i, j such that $U_i \cap U_j$ is not empty we are given an r-form ω_{ij} on $U_i \cap U_j$; and suppose that for each triple of subscripts i, j, k such that $U_i \cap U_j \cap U_k$ is not empty these forms satisfy

$$\omega_{ij} + \omega_{jk} + \omega_{ki} = 0 \text{ in } U_i \cap U_j \cap U_k. \tag{26}$$

Then there is an r-form ω_i on U_i for each i such that

$$\omega_{ij} = \omega_i - \omega_j \text{ on } U_i \cap U_j \tag{27}$$

whenever $U_i \cap U_j$ is not empty.

Proof. From (26) we deduce first $\omega_{ii} = 0$ and then $\omega_{ji} = -\omega_{ij}$. We now construct a partition of unity corresponding to the covering $\{U_1, \ldots, U_N\}$ - that is, a set of infinitely differentiable functions $f_1, \ldots, f_N$ on T such that

(i) $f_\nu \geq 0$ in U_ν and $f_\nu = 0$ outside U_ν for each ν, and

(ii) $f_1 + \ldots + f_N = 1$ at each point of T.

For this purpose, associate with U_ν real local coordinates $\xi_1, \ldots, \xi_n, \eta_1, \ldots, \eta_n$ such that U_ν is the set of points satisfying

$$\xi_1^2 + \eta_1^2 + \ldots + \xi_n^2 + \eta_n^2 < c^2$$

for some constant c; that this is possible follows at once from the definition of an open hypersphere in T given at the end of §3. Define the function ϕ_ν on T by

$$\phi_\nu = \exp\{-1/(c^2 - \xi_1^2 - \eta_1^2 - \ldots - \xi_n^2 - \eta_n^2)\} \text{ in } U_\nu$$

and $\phi_\nu = 0$ outside U_ν. Clearly $\phi_\nu > 0$ in U_ν and ϕ_ν is infinitely differentiable everywhere on T. Now write

$$f_\nu = \phi_\nu/(\phi_1 + \ldots + \phi_N) \text{ for } \nu = 1, 2, \ldots, N;$$

then the f_ν have the required properties. Let

$$\omega_i = f_1\omega_{i1} + \ldots + f_N\omega_{iN} \quad \text{in } U_i$$

where any term which is undefined at a point is to be taken to be zero at that point. It is easy to check that ω_i is an r-form (that is to say, the functions that occur in it are infinitely differentiable) and (27) holds.

Lemma 17. Let ω be a 2-form on $T = V/\Lambda$ such that $d\omega = 0$, let ω^* be the induced form on V and let $\xi_1, \ldots, \xi_{2n}$ be real coordinates on V. Then there is a 1-form ψ on T, with induced form ψ^* on V, and a 1-form $\phi = \Sigma\Sigma c_{ij}\xi_i d\xi_j$ on V, where the c_{ij} are constants, such that $\omega^* = d\psi^* + d\phi$.

Proof. The truth of the Lemma is not affected by a change of coordinates, so we may assume that Λ consists of all points with integer coordinates. We have

$$\omega^* = \Sigma\Sigma f_{ij} d\xi_i \wedge d\xi_j$$

where the f_{ij} are periodic and infinitely differentiable; so each of them can be expanded as a multiple Fourier series

$$f_{ij} = \Sigma\, a_{ij}^{(m)} \exp\{2\pi i(m_1\xi_1 + \ldots + m_{2n}\xi_{2n})\} \tag{28}$$

where the sum is over all integer vectors m. Since the exponential factors are not altered by differentiation, terms with different exponentials cannot interact; so for each integer vector m we have

$$d[\text{multiple of } \exp\{2\pi i\Sigma m_\nu\xi_\nu\} \text{ in } \omega^*] = 0.$$

To prove the Lemma we must therefore prove first that it holds for the individual components of ω^*, that is in the special case

$$\omega^* = \{\exp(2\pi i\Sigma m_\nu\xi_\nu)\}\Sigma\Sigma a_{ij} d\xi_i \wedge d\xi_j, \tag{29}$$

and second that in the general case the resulting formal Fourier series for ψ^* is a 1-form - that is, the functions that occur in it are infinitely differentiable.

In the case $m = 0$ we can take $\psi^* = 0$ and $\phi = \Sigma\Sigma a_{ij}\xi_i d\xi_j$; so henceforth we may assume $m \neq 0$ and so without loss of generality

$m_1 \neq 0$. Moreover in (29) we may assume that $a_{ij} = 0$ unless $i < j$. Now by considering the term in $d\xi_1 \wedge d\xi_i \wedge d\xi_j$ in $d\omega^* = 0$ we obtain

$$m_1 a_{ij} - m_i a_{1j} + m_j a_{1i} = 0 \quad \text{for } 1 < i < j;$$

and it is now easy to verify that

$$\omega^* = (2\pi i)^{-1} d\{m_1^{-1} \exp(2\pi i \sum_1^{2n} m_\nu \xi_\nu) \sum_2^{2n} a_{1j} d\xi_j\}.$$

Thus $\omega^* = d\psi^*$ for the obvious 1-form ψ.

It is well known that the series (28) represents an infinitely differentiable function if and only if

$$a_{ij}^{(m)} = O((1 + |m_1| + \ldots + |m_{2n}|)^{-N})$$

for each integer N. In going from the Fourier series for ω^* to that for ψ^* we have made the coefficients smaller; so the formal Fourier series for ψ^* represents a 1-form, and this completes the proof of the Lemma.

Corollary. If ω contains no terms of type (0, 2) then we can further require that ϕ and ψ are of type (1, 0).

Proof. In the proof of the Lemma we had to choose ϕ so that $d\phi$ was equal to the constant term in the Fourier expansion of ω^*. With the extra hypothesis of the Corollary, this means we must choose ϕ to satisfy

$$d\phi = \Sigma\Sigma\alpha_{ij} dz_i \wedge dz_j + \Sigma\Sigma\beta_{ij} dz_i \wedge d\bar{z}_j$$

for given constants α_{ij}, β_{ij}; and for this we may take

$$\phi = \Sigma\Sigma\alpha_{ij} z_i dz_j - \Sigma\Sigma\beta_{ij} \bar{z}_j dz_i$$

which is of type (1, 0). Now let

$$\exp\{2\pi i \Sigma(\alpha_\nu z_\nu + \bar{\alpha}_\nu \bar{z}_\nu)\} \Sigma(\beta_j dz_j + \gamma_j d\bar{z}_j) \qquad (30)$$

be a general term in the Fourier expansion of the ψ^* obtained in the proof of Lemma 17. Since this ψ^* had no constant term, some $\alpha_\nu \neq 0$ and without loss of generality we can assume $\alpha_1 \neq 0$. Since the derivative of (30) has no terms of type (0, 2), we have $\bar{\alpha}_1\gamma_j = \bar{\alpha}_j\gamma_1$; and thus if we subtract

$$d[(\gamma_1/2\pi i\bar{\alpha}_1)\exp\{2\pi i\Sigma(\alpha_\nu z_\nu + \bar{\alpha}_\nu\bar{z}_\nu)\}]$$

from (30), which does not alter its derivative, we obtain a form of type (1, 0). It is easy to check that this process at most doubles the coefficients in (30), so the modified ψ still has the differentiability properties required.

Theorem 18. Let $\mathfrak{a}$ be a positive divisor on T regarded as a complex torus, and let $\mathfrak{a}^*$ be the induced divisor on V; then there is a theta-function with respect to Λ whose divisor is $\mathfrak{a}^*$.

Proof. By Lemma 14 we can assume that $\mathfrak{a}$ has a finite description $\{U_i, \phi_i\}$ in which each U_i is an open hypersphere in T. Thus ϕ_i/ϕ_j is holomorphic and nowhere zero on $U_i \cap U_j$ and so the

$$\omega_{ij} = d\log(\phi_i/\phi_j) \quad \text{on } U_i \cap U_j$$

are forms of type (1, 0) which satisfy the hypothesis (26) of Lemma 16. Thus there are 1-forms ω_i on U_i such that

$$\omega_i - \omega_j = \omega_{ij} = d\log(\phi_i/\phi_j) \quad \text{on } U_i \cap U_j. \tag{31}$$

Moreover we can suppose the ω_i all to be of type (1, 0), either by looking at the proof of Lemma 16 or by deleting the terms of type (0, 1). Now (31) gives $d\omega_i = d\omega_j$ on $U_i \cap U_j$, so the $d\omega_i$ piece together to give a 2-form ω on T which has no terms of type (0, 2). Locally $d\omega = d(d\omega_i) = 0$; so we can write $\omega^* = d\psi^* + d\phi$ where ϕ and ψ have all the properties stated in Lemma 17 and its Corollary. Now write $\psi_i = \omega_i - \psi$ on U_i; thus the ψ_i are forms of type (1, 0) such that

$$d\log(\phi_i/\phi_j) = \psi_i - \psi_j \quad \text{on } U_i \cap U_j.$$

The inverse image of U_i in V is the disjoint union of open hyperspheres $U_{i\lambda}$ where λ runs through the elements of Λ and the notation is so chosen that $U_{i\lambda}$ is the translation of U_{i0} by λ. Moreover $d(\psi_i^* - \phi) = 0$

on $U_{i\lambda}$ and hence by Lemma 15 there is a function $f_{i\lambda}$ on $U_{i\lambda}$ such that

$$df_{i\lambda} = \psi_i^* - \phi \quad \text{on } U_{i\lambda}.$$

Since the right hand side is of type (1, 0), $f_{i\lambda}$ is holomorphic on $U_{i\lambda}$; and

$$df_{i\lambda} - df_{j\mu} = \psi_i^* - \psi_j^* = d\log(\phi_i/\phi_j)^* \quad \text{on } U_{i\lambda} \cap U_{j\mu}$$

provided that this intersection is not empty. Hence $\phi_i^* \exp(-f_{i\lambda})$ is a constant multiple of $\phi_j^* \exp(-f_{j\mu})$ in this intersection, both being holomorphic there; and in this way we can analytically continue $\phi_1^* \exp(-f_{10})$ say to a function $\theta(z)$ holomorphic in the whole of V. Clearly the divisor of $\theta(z)$ is $\mathfrak{a}^*$, so it only remains to prove that $\theta(z)$ is a theta-function for Λ. Fix λ in Λ and suppose say that z is in U_{10}; then $z + \lambda$ is in $U_{1\lambda}$ and by the periodicity of ϕ_1^*,

$$\theta(z + \lambda)/\theta(z) = c \exp\{f_{10}(z) - f_{1\lambda}(z + \lambda)\}$$

for some constant $c \neq 0$. Writing

$$\phi = \Sigma\Sigma(a_{ij}z_i + b_{ij}\bar{z}_i)dz_j$$

we obtain

$$\begin{aligned} d\log\{\theta(z + \lambda)/\theta(z)\} &= df_{10}(z) - df_{1\lambda}(z + \lambda) \\ &= \phi(z + \lambda) - \phi(z) = \Sigma\Sigma(a_{ij}\lambda_i + b_{ij}\bar{\lambda}_i)dz_j \end{aligned}$$

where exceptionally the λ_i are the complex coordinates of λ. The functional equation for $\theta(z + \lambda)/\theta(z)$ follows by integration, and this completes the proof of the Theorem. By definition, $\mathfrak{a}$ determines $\theta(z)$ up to multiplication by a trivial theta-function.

Corollary. Any multiply periodic function can be written as a quotient of theta functions.

Proof. Let f be multiply periodic and not identically zero; then by Theorem 13 we can write $(f) = \mathfrak{a}_1^* - \mathfrak{a}_2^*$ where $\mathfrak{a}_1$ and $\mathfrak{a}_2$ are positive divisors on T. Let θ be a theta-function whose divisor is $\mathfrak{a}_2^*$; then $f\theta$ is holomorphic because its divisor is $\mathfrak{a}_1^* \geq 0$, and it satisfies the same functional equation as θ. So $f\theta$ is a theta-function and this proves the Corollary.

5. Consistency conditions and Riemann forms

In the functional equations

$$\theta(z + \lambda) = \theta(z)\exp F(z, \lambda) \tag{32}$$

for a theta function which does not vanish identically, the $F(z, \lambda)$ for various λ cannot be unrelated. In this section we first show how to describe $F(z, \lambda)$ in terms of the $F(z, \lambda_\nu)$ where $\lambda_1, \ldots, \lambda_{2n}$ is a base for Λ, and then we find necessary conditions on the $F(z, \lambda)$ for (32) to have a non-zero solution. In §6 we show that these conditions are also sufficient, and determine in terms of the $F(z, \lambda)$ the dimension of the linear space of solutions of (32).

Now let λ, λ' be any elements of Λ. The identity

$$\frac{\theta(z+\lambda+\lambda')}{\theta(z)} = \frac{\theta(z+\lambda+\lambda')}{\theta(z+\lambda)} \cdot \frac{\theta(z+\lambda)}{\theta(z)}$$

together with (32) gives

$$F(z, \lambda + \lambda') \equiv F(z + \lambda, \lambda') + F(z, \lambda) \bmod 2\pi i. \tag{33}$$

It is now convenient to split F into its linear and constant parts, writing

$$F(z, \lambda) = 2\pi i \{L(z, \lambda) + J(\lambda)\}$$

where $L(z, \lambda)$ is homogeneous **C**-linear in its first argument; thus $L(z, \lambda)$ is fully determined but $J(\lambda)$ is only determined mod 1. When we substitute this into (33) we can separate the constant terms from the linear terms by setting $z = 0$, and since the congruence for the linear terms can only hold if it is an identity, we obtain

$$L(z, \lambda + \lambda') = L(z, \lambda) + L(z, \lambda'), \tag{34}$$

$$J(\lambda + \lambda') - J(\lambda) - J(\lambda') \equiv L(\lambda, \lambda') \bmod 1. \tag{35}$$

Since $V = \Lambda \otimes \mathbf{R}$, (34) shows that L can be uniquely extended to a $\mathbf{C}$-valued function on $V \times V$ which is $\mathbf{R}$-linear in the second argument and still $\mathbf{C}$-linear in the first. Denote the extended function also by L, and write

$$E(z, w) = L(z, w) - L(w, z), \tag{36}$$

$$H(z, w) = E(iz, w) + iE(z, w). \tag{37}$$

Theorem 19. E is $\mathbf{R}$-bilinear, real-valued and alternating on $V \times V$; and $E(z, w) = E(iz, iw)$. $S(z, w) = E(iz, w)$ is $\mathbf{R}$-bilinear, real-valued and symmetric; and $S(z, w) = S(iz, iw)$. H is Hermitian. Moreover E is integer-valued on $\Lambda \times \Lambda$.

Proof. E is $\mathbf{R}$-bilinear and alternating by (36). Since the left hand side of (35) is symmetric in λ and λ', $L(\lambda, \lambda') \equiv L(\lambda', \lambda) \bmod 1$ and so E is integer-valued on $\Lambda \times \Lambda$; since it is $\mathbf{R}$-bilinear, it is also real-valued on $V \times V$. Moreover

$$\begin{aligned} E(z, w) - E(iz, iw) &= L(z, w) + L(iw, iz) - L(w, z) - L(iz, iw) \\ &= i\{E(w, iz) - E(z, iw)\} \end{aligned}$$

since L is $\mathbf{C}$-linear in its first argument. Since the first of these expressions is real and the last is purely imaginary, they must both vanish; and with a little manipulation this completes the proof of the first sentence of the Theorem. But an easy calculation using (37) shows that the second and third sentences follow from the first; indeed given (37) and the fact that E is the imaginary part of H, any two of the first three sentences are equivalent. This completes the proof of the Theorem.

A function E which satisfies the conditions of Theorem 19 is called an alternating Riemann form on $V \times V$ with respect to Λ, and the corresponding H is called a Hermitian Riemann form. It is worth while to consider the interpretation of Theorem 19 in the special case $n = 1$, and to compare it with the results of §2; so let $n = 1$ and let λ_1, λ_2 be a base for Λ such that $\mathrm{Im}(\lambda_1/\lambda_2) > 0$. If Γ denotes as before the boundary of the fundamental parallelogram then by (32) and the same argument as in the proof of (22),

$$\frac{1}{2\pi i}\int_\Gamma \frac{\theta'(z)}{\theta(z)}\,dz = L(\lambda_1, \lambda_2) - L(\lambda_2, \lambda_1) = m$$

where m is the number of zeros of θ in a period parallelogram. This gives the interpretation of $E(\lambda_1, \lambda_2)$ as an integer and indeed shows that it must be positive; the generalization of this last fact will appear in §9. Since E is $\mathbf{R}$-bilinear and alternating, it is completely determined by a knowledge of $E(\lambda_1, \lambda_2)$. Moreover $E(z, w)$ must be a constant multiple of the area of the (oriented) parallelogram generated by Oz and Ow; so the remaining condition $E(z, w) = E(iz, iw)$ is satisfied automatically. Thus to given Λ and m there corresponds just one Riemann form; it is for this reason that the Riemann form did not appear explicitly in §2.

When $n > 1$, however, a general lattice Λ will admit no non-zero Riemann forms. For if $\lambda_1, \ldots, \lambda_{2n}$ is a base for Λ then E as an $\mathbf{R}$-bilinear alternating form is uniquely determined by the $E(\lambda_i, \lambda_j)$, which are integers; and the condition $E(z, w) = E(iz, iw)$ induces linear relations with real coefficients between the $E(\lambda_i, \lambda_j)$, which for general Λ have no non-trivial integer solutions. However it can be shown that with the obvious topology the set of lattices Λ (for fixed $n > 1$) which admit non-zero Riemann forms is dense in the set of all lattices.

We now turn our attention to (35). Writing

$$J(\lambda) = K(\lambda) + \tfrac{1}{2}L(\lambda, \lambda)$$

in (35), we obtain

$$K(\lambda + \lambda') - K(\lambda) - K(\lambda') \equiv \tfrac{1}{2}E(\lambda, \lambda') \bmod 1. \tag{38}$$

Now let $B(z, w)$ be an $\mathbf{R}$-bilinear form on $V \times V$ which is integer-valued on $\Lambda \times \Lambda$ and satisfies

$$E(z, w) = B(z, w) - B(w, z);$$

that such B exist can easily be shown by considering the matrix representation of E with respect to a base for Λ. Write

$$K(\lambda) = K'(\lambda) + \tfrac{1}{2}B(\lambda, \lambda);$$

then (38) reduces to

$$K'(\lambda + \lambda') \equiv K'(\lambda) + K'(\lambda') \bmod 1,$$

so that (35) is equivalent to the statement that K' is a homomorphism $\Lambda \to \mathbf{C}/\mathbf{Z}$.

We say that two theta-functions are equivalent if their quotient is a trivial theta-function; thus by Theorem 18 each positive divisor on V/Λ corresponds to an equivalence class of theta-functions. Suppose that in the calculations above we replace $\theta(z)$ by

$$\theta'(z) = \theta(z)\exp\{2\pi i[Q(z, z) + R(z) + S]\} \tag{39}$$

where Q is $\mathbf{C}$-bilinear symmetric, R is $\mathbf{C}$-linear and S is constant; thus $\theta'(z)$ is the most general theta-function equivalent to $\theta(z)$. Then $L(z, \lambda)$ is increased by $2Q(z, \lambda)$ and $J(\lambda)$ is increased by $Q(\lambda, \lambda) + R(\lambda)$; thus $E(z, w)$ remains unchanged and $K(\lambda)$ is increased by $R(\lambda)$, as also is $K'(\lambda)$. But if E is given, (36) determines L up to addition of a $\mathbf{C}$-bilinear symmetric function; thus within an equivalence class of theta-functions we can obtain all L compatible with (36). We can use this in various ways to pick out from each equivalence class a normalized theta-function, determined up to multiplication by a constant. For example

$$\tfrac{1}{2}iH(w, z) - \tfrac{1}{2}iH(z, w) = E(z, w)$$

so that to choose $L(z, w) = -\tfrac{1}{2}iH(z, w)$ is compatible with (36). Again, let $f(\lambda)$ be the imaginary part of $K(\lambda)$; then (38) shows that $f(\lambda + \lambda') = f(\lambda) + f(\lambda')$, so that f can be uniquely extended to an $\mathbf{R}$-linear real-valued function on V. Take

$$R(z) = -f(iz) - if(z);$$

then $K(\lambda) + R(\lambda)$ is real-valued and $R(z)$ is $\mathbf{C}$-linear. We shall say that $\theta(z)$ is a normalized theta-function if $L(z, w) = -\tfrac{1}{2}iH(z, w)$ and $K(\lambda)$ and $K'(\lambda)$ are real-valued.

Lemma 20. Every equivalence class of theta-functions contains a normalized theta-function, which is unique up to multiplication by a constant.

We have already proved existence. As to uniqueness, suppose that θ and θ' in (39) are both normalized; then $Q = 0$ and $R(z)$ is real-valued and hence must vanish.

Since E is the same for all theta-functions in an equivalence class, it is determined by knowing the divisor of $\theta(z)$; and in view of Theorem 18 we can associate to each positive divisor on V/Λ a Riemann form E. This map is easily seen to be additive, and therefore extends to a homomorphism from the group of divisors on V/Λ to the group of Riemann forms with respect to Λ. In particular, to a trivial theta-function corresponds the trivial Riemann form $E = 0$. Conversely we shall see below that if a holomorphic theta-function has $E = 0$ then it is trivial; in other words the kernel of the homomorphism above contains no non-trivial positive divisors. A fuller investigation of the kernel will be given in §8. The homomorphism is not necessarily onto; examples where it is not onto can be constructed with the help of the next Lemma.

Lemma 21. The Hermitian Riemann form H associated with a theta-function $\theta(z)$ is positive semi-definite; that is, $H(z, z) \geq 0$ for all z in V.

Proof. Without loss of generality we can assume that $\theta(z)$ is normalized. Write

$$\phi(z) = \theta(z)\exp\{-\tfrac{1}{2}\pi H(z, z)\}$$

so that $\phi(z)$ is continuous although not holomorphic. We have

$$\theta(z + \lambda) = \theta(z)\exp\{\pi H(z, \lambda) + \tfrac{1}{2}\pi H(\lambda, \lambda) + 2\pi i K(\lambda)\} \qquad (40)$$

and so the functional equation of $\phi(z)$ is

$$\phi(z + \lambda) = \phi(z)\exp\{\pi i[E(z, \lambda) + 2K(\lambda)]\}.$$

Here the expression in square brackets is real, so that $|\phi(z)|$ is periodic and therefore bounded; thus there is a constant A such that

$$|\theta(z)| \leq A \exp\{\tfrac{1}{2}\pi H(z, z)\}. \qquad (41)$$

Now suppose that there exists z_0 such that $H(z_0, z_0) < 0$, and set $z = tz_0$ with t in **C**. Viewed as a function of t, $\theta(tz_0)$ is holomorphic; and by (41) it tends to 0 as $|t| \to \infty$. So $\theta(tz_0)$ vanishes identically, and in particular $\theta(z_0) = 0$. But $H(z, z) < 0$ for all z in some neighbourhood of z_0; so by the same argument $\theta(z) = 0$ in that neighbourhood, and hence everywhere by analytic continuation. This is absurd; so there is no such z_0 and the Lemma is proved.

We say that a theta-function if <u>degenerate</u> if the corresponding H is not positive definite; we now show that a degenerate theta-function is essentially the same as a theta-function on a quotient space of V. For any positive semi-definite Hermitian Riemann form H we define the <u>kernel</u> of H to be the set of w such that $H(z, w) = 0$ for all z in V; clearly the kernel is a **C**-vector space.

Lemma 22. <u>Let H be a positive semi-definite Hermitian Riemann form on V, and let W be its kernel. Then W consists of all w such that $H(w, w) = 0$; and H induces a positive definite Hermitian form on V/W. Moreover the image of Λ in V/W is a lattice and the induced form is a Riemann form with respect to that lattice.</u>

Proof. Let w be such that $H(w, w) = 0$. For any t in **C** and z in V,

$$0 \leq H(tw + z, tw + z) = 2\mathrm{Re}\{tH(w, z)\} + H(z, z).$$

For fixed z, this inequality can only hold for all t if $H(w, z) = 0$; and since this must hold for all z, w must be in W. The converse is trivial. Moreover $H(w + z, w' + z') = H(z, z')$ for any w, w' in W and z, z' in V, so that H induces a bilinear form on V/W; and it is easy to see that this form is Hermitian and positive definite. The image of Λ in V/W clearly spans V/W, so to prove it is a lattice we need only show that it is discrete. Fix a base $\lambda_1, \ldots, \lambda_{2n}$ for Λ and let N be a neighbourhood of the origin in V/W such that

$$|H(z, \lambda_\nu)| < 1 \quad \text{for } \nu = 1, 2, \ldots, 2n$$

whenever the image of z is in N; such an N exists because H induces a form on V/W. Let λ be a point of Λ whose image is in N; then $|E(\lambda, \lambda_\nu)| < 1$ for each ν since E is the imaginary part of H, and so $E(\lambda, \lambda_\nu) = 0$ since it must be integral. This implies $E(\lambda, z) = 0$ for all z in C, whence $H(\lambda, z) = 0$ for all z; so λ is in W and its image in V/W is the origin. Thus the image of Λ in V/W is discrete and the Lemma follows at once.

Theorem 23. Let $\theta(z)$ be a normalized theta-function on V with respect to Λ, let H be the associated Hermitian Riemann form and let W be the kernel of H. If Λ' is the image of Λ in V/W then θ induces a function θ' on V/W which is a theta-function with respect to Λ', and the Hermitian Riemann form on V/W corresponding to θ' is the form on V/W induced by H.

Proof. We need only prove that $\theta(z)$ induces a function on V/W - that is, that $\theta(z)$ is constant on each coset of W - since the rest of the Theorem will then follow trivially. Suppose that z is in V and w in W; then (41) implies

$$|\theta(z + w)| \leq A \exp\{\tfrac{1}{2}\pi H(z, z)\}$$

for a constant A which depends only on θ. Thus $\theta(z + w)$, regarded as a function of w for fixed z, is holomorphic and bounded; so it is constant and in particular $\theta(z + w) = \theta(z)$. This is what we needed to prove. The condition that $\theta(z)$ is normalized is essential; but it follows that any degenerate theta-function can be built up from a trivial theta-function and a non-degenerate theta-function on a proper quotient space.

Corollary. If $H = 0$ then $\theta(z)$ is trivial.

For now $W = V$ and so the normalized theta-function is constant.

Now suppose that H is non-degenerate or, which is equivalent, that E is non-singular. As a **Z**-valued alternating form on $\Lambda \times \Lambda$, E is represented by a non-singular skew-symmetric matrix with elements in **Z** as soon as we have chosen a base for Λ; and the determinant of this matrix is a perfect square which does not depend on the choice of base. The positive square root of this determinant is called the Pfaffian of E. By suitable choice of a base for Λ, we can even arrange that the matrix for E has the form

$$\begin{pmatrix} 0 & D \\ -D & 0 \end{pmatrix} \text{ where } D = \operatorname{diag}(d_1, \ldots, d_n)$$

and the d_ν are positive integers such that d_ν divides $d_{\nu+1}$ for $\nu = 1, 2, \ldots, n-1$. The Pfaffian of E is then equal to $d_1 d_2 \ldots d_n$. The d_ν are uniquely determined by E, even though the corresponding base for Λ is not unique.

A theta-function for Λ will also be a theta-function for any sub-lattice $\Lambda_1 \subset \Lambda$, and its Pfaffian with respect to Λ_1 will be $[\Lambda : \Lambda_1]$ times its Pfaffian with respect to Λ. It may also be a theta-function with respect to some $\Lambda_2 \supset \Lambda$; but if it is non-degenerate there can only be finitely many such Λ_2 because $[\Lambda_2 : \Lambda]$ must be a factor of its Pfaffian with respect to Λ.

6. Construction of theta functions

To find all the theta-functions corresponding to given L and J, we essentially mimic the construction at the end of §2. There is however one extra complication. The naive way to proceed would be to take any base $\lambda_1, \ldots, \lambda_{2n}$ for Λ such that $\lambda_1, \ldots, \lambda_n$ are linearly independent over **C**, and then multiply the theta-functions we are looking for by an assigned trivial theta-function so chosen that the product is periodic with periods $\lambda_1, \ldots, \lambda_n$. However when $n > 1$ there are not enough free parameters in the general trivial theta-function to do this in a straightforward way, and it needs a special choice of $\lambda_1, \ldots, \lambda_n$ to make it possible. This is the reason for the opening steps of the proof of Theorem 24 below.

Clearly we can assume that L and J satisfy (34) and (35), and that the Riemann forms E and H have the properties stated in Theorem 19 and Lemma 21. Moreover Theorem 23 enables us to reduce to the case when H is positive definite.

Theorem 24. <u>Suppose that</u> $L(z, w)$ <u>is</u> **C**-<u>linear in its first argument and</u> **R**-<u>linear in its second, that</u> $J(\lambda)$ <u>satisfies (35), that</u> E <u>and</u> H <u>defined by (36) and (37) have the properties stated in Theorem 19, and that</u> $H(z, z)$ <u>is positive definite. Then the space of functions</u> $\theta(z)$ <u>holomorphic on</u> V <u>and satisfying</u>

$$\theta(z + \lambda) = \theta(z)\exp\{2\pi i[L(z, \lambda) + J(\lambda)]\} \tag{42}$$

<u>for each</u> λ <u>in</u> Λ <u>has dimension equal to the Pfaffian of</u> E.

Proof. As at the end of §5, we can choose a base $\lambda_1, \ldots, \lambda_{2n}$ for Λ with respect to which E is represented by the matrix $\begin{pmatrix} 0 & D \\ -D & 0 \end{pmatrix}$ where $D = \operatorname{diag}(d_1, \ldots, d_n)$ and the d_ν are positive integers. Suppose that $\lambda_1, \ldots, \lambda_n$ were linearly dependent over **C**, so that there was a relation

$$(a_1 + ib_1)\lambda_1 + \ldots + (a_n + ib_n)\lambda_n = 0$$

with the a_ν, b_ν real and not all zero. Thus we would have

$$\Sigma\, a_\nu \lambda_\nu = \alpha, \quad \Sigma\, b_\nu \lambda_\nu = -i\alpha$$

for some α in V; and $\alpha \neq 0$ since $\lambda_1, \ldots, \lambda_n$ are linearly independent over **R**. But now

$$0 < H(\alpha, \alpha) = E(i\alpha, \alpha) = \Sigma\Sigma\, a_\mu b_\nu E(\lambda_\mu, \lambda_\nu) = 0$$

by the choice of the matrix representation of E. This is absurd; so $\lambda_1, \ldots, \lambda_n$ are linearly independent over **C** and we can take them to be a base for the system of coordinates $z_1, \ldots, z_n$ on V.

By multiplying $\theta(z)$ by $\exp\{2\pi i[Q(z, z) + R(z)]\}$, where Q is **C**-bilinear symmetric and R is **C**-linear, we increase $L(z, w)$ by $2Q(z, w)$ and $K(\lambda)$ by $R(\lambda)$. Choose

$$Q(z, w) = -\tfrac{1}{2}\Sigma\Sigma\, z_\mu w_\nu L(\lambda_\mu, \lambda_\nu), \quad R(z) = -\Sigma\, z_\nu K(\lambda_\nu)$$

where all sums run from 1 to n, and replace the old L, J, K by the new; since L is **C**-linear in its first argument we obtain for $\nu = 1, 2, \ldots, n$

$$L(z, \lambda_\nu) = 0, \quad K(\lambda_\nu) = 0 \tag{43}$$

and thus also $J(\lambda_\nu) = 0$. In other words, the modified $\theta(z)$ has period 1 in each of $z_1, \ldots, z_n$ and can therefore be written as a multiple Fourier series

$$\theta(z) = \Sigma \ldots \Sigma\, c(m_1, \ldots, m_n)\exp\{2\pi i(m_1 z_1 + \ldots + m_n z_n)\}.$$

Since (34) and (35) are satisfied, the arguments at the beginning of §5 show that (42) holds for all λ in Λ if it holds for $\lambda = \lambda_1, \ldots, \lambda_{2n}$. The first n of these cases have been dealt with already. Moreover for μ, ν between 1 and n

$$L(\lambda_\mu, \lambda_{n+\nu}) = -E(\lambda_{n+\nu}, \lambda_\mu) = \begin{cases} d_\nu & \text{if } \mu = \nu \\ 0 & \text{otherwise} \end{cases}$$

by (43); thus $L(z, \lambda_{n+\nu}) = d_\nu z_\nu$ and in particular the νth coordinate of $\lambda_{n+\mu}$ is equal to $d_\nu^{-1} L(\lambda_{n+\mu}, \lambda_{n+\nu})$. The functional equation (42) with $\lambda = \lambda_{n+\nu}$ now takes the form

$$\theta(z + \lambda_{n+\nu}) = \theta(z)\exp\{2\pi i[d_\nu z_\nu + J(\lambda_{n+\nu})]\}$$

which is equivalent to the set of equations

$$\begin{aligned} c(m_1, \ldots, m_\nu - d_\nu, \ldots, m_n)/c(m_1, \ldots, m_n) \\ = \exp\{2\pi i[\Sigma\, m_\mu d_\mu^{-1} L(\lambda_{n+\nu}, \lambda_{n+\mu}) - J(\lambda_{n+\nu})]\}. \end{aligned}$$

Formally this means that we can choose the $c(m_1, \ldots, m_n)$ with $0 \le m_\nu < d_\nu$ for each ν and all the other coefficients are determined by the last equation. This gives $d_1 d_2 \ldots d_n$ degrees of freedom, and $d_1 d_2 \ldots d_n$ is just the Pfaffian of E. It only remains to check that the resulting formal Fourier series converge. But it is easy to see that provided each m_ν stays in a fixed congruence class mod d_ν,

$$c(m_1, \ldots, m_n) = \exp\{-\pi i L(\beta, \beta) + \text{linear form in the } m_\nu + \text{constant}\}$$

where $\beta = \Sigma\, m_\nu d_\nu^{-1} \lambda_{n+\nu}$; so to prove convergence it is sufficient to show that the imaginary part of $L(z, z)$ is negative definite. Write $z = x + iy$ where x and y are linear combinations of $\lambda_1, \ldots, \lambda_n$ with real coefficients; then

$$L(z, z) = L(x, z) + iL(y, z) = E(x, z) + iE(y, z)$$

by (43). The imaginary part of this is

$$E(y, z) = E(y, x) + E(y, iy) = -H(y, y) < 0$$

for $y \neq 0$; and since here z lies in the real vector space spanned by $\lambda_{n+1}, \ldots, \lambda_{2n}$ we cannot have $z \neq 0$ and $y = 0$, for this would imply a linear dependence relation over $\mathbf{R}$ between $\lambda_1, \ldots, \lambda_{2n}$. This completes the proof of the Theorem.

Corollary. Under the conditions of the Theorem, almost all of the theta-functions satisfying (42) are not theta-functions for any lattice Λ' strictly containing Λ.

Proof. Let $m = [\Lambda' : \Lambda]$ and let d be the Pfaffian of E with respect to Λ. The vector space of solutions of (42) has dimension d. It was shown at the end of §5 that if there is a solution of (42) which is a theta-function with respect to Λ' then m divides d; so there are only finitely many candidates for Λ'. To each of them corresponds only finitely many ways of extending the functional equation (42) to Λ'; indeed L extends in just one way and K' in just m ways. To each of these extensions corresponds a space of solutions of dimension d/m; so the solutions which are theta-functions for some Λ' form a finite set of proper subspaces, and this proves the Corollary.

We shall say that two divisors $\mathfrak{a}_1$ and $\mathfrak{a}_2$ on $T = V/\Lambda$ are linearly equivalent if their difference is the divisor of a meromorphic function on T; and we write this as $\mathfrak{a}_1 \sim \mathfrak{a}_2$. Clearly this is an equivalence relation and is compatible with addition and subtraction of divisors; moreover if two divisors are linearly equivalent their associated Riemann forms are equal. We shall say that a positive divisor is degenerate if the corresponding Riemann form is degenerate; after Theorem 23 this happens if and only if there are infinitesimal translations under which the divisor is invariant. By analogy with §1, we shall for any divisor $\mathfrak{a}$ on T define $L(\mathfrak{a})$ to be the $\mathbf{C}$-vector space of those meromorphic functions f on T for which $(f) + \mathfrak{a} \geq 0$, together with the zero function; and we write $l(\mathfrak{a}) = \dim L(\mathfrak{a})$.

Lemma 25. Let $\mathfrak{a}_0, \ldots, \mathfrak{a}_m$ be positive divisors on T and suppose that $\mathfrak{a}_0$ at least is non-degenerate. Then there is a homogeneous polynomial $P(r_0, \ldots, r_m)$ of degree n such that

$$l(r_0\mathfrak{a}_0 + \ldots + r_m\mathfrak{a}_m) = P(r_0, \ldots, r_m)$$

whenever r_0 is positive and $r_1, \ldots, r_m$ are non-negative.

Proof. Let E_μ be the Riemann form corresponding to $\mathfrak{a}_\mu$; then the form corresponding to $r_0\mathfrak{a}_0 + \ldots + r_m\mathfrak{a}_m$ will be $r_0E_0 + \ldots + r_mE_m$, which is non-singular because the associated Hermitian form is positive definite. The determinant of $r_0E_0 + \ldots + r_mE_m$ with respect to Λ is a homogeneous polynomial of degree 2n and is a perfect square for all allowable values of the r_μ; it follows easily that it is a square when viewed as a polynomial, and hence the Pfaffian is a polynomial of degree n in the r_μ. The Lemma now follows from Theorem 24.

Lemma 26. Suppose that there is a non-degenerate positive divisor on T; then the field of meromorphic functions on T has transcendence degree at most n over **C**, and if it has transcendence degree exactly n then it is finitely generated over **C**.

Proof. Let $f_1, \ldots, f_{n+1}$ be meromorphic functions on T, and let $\mathfrak{a}$ be a positive non-degenerate divisor on T such that $(f_\nu) + \mathfrak{a} \geq 0$ for each ν. The number of distinct monomials in the f_ν of total degree at most N is

$$(N + n + 1)!/N!(n + 1)! = N^{n+1}/(n + 1)! + O(N^n).$$

All these monomials lie in $L(N\mathfrak{a})$, whose dimension is $O(N^n)$ by Lemma 25. So when N is large enough there is a linear dependence relation between the monomials, and this is an algebraic relation among $f_1, \ldots, f_{n+1}$.

Now suppose that $f_1, \ldots, f_n$ are meromorphic functions on T algebraically independent over **C**; let $\mathfrak{a}_0$ be a positive non-degenerate divisor on T such that $(f_\nu) + \mathfrak{a}_0 \geq 0$ for each ν, let d_0 be the Pfaffian of the Riemann form associated with $\mathfrak{a}_0$, and write $M = n!\,d_0$. Let g be any meromorphic function on T, and let $\mathfrak{a}_1$ be a positive divisor on T such that $(g) + \mathfrak{a}_1 \geq 0$. By Lemma 25 and Theorem 24,

$$l(N\mathfrak{a}_0 + M\mathfrak{a}_1) = d_0 N^n + O(N^{n-1}) \quad \text{as } N \to \infty.$$

On the other hand the corresponding linear space contains all the monomials

$$f_1^{m_1} \ldots f_n^{m_n} g^m$$

with $m_1, \ldots, m_n$ non-negative, $m_1 + \ldots + m_n \leq N$ and $0 \leq m \leq M$. The number of these is

$$(N + n)!\,(M + 1)/N!\,n! = (d_0 + 1/n!)N^n + O(N^{n-1});$$

so when N is large enough there is a linear dependence relation between them. This relation must really involve g, because the f_ν are algebraically independent; so it shows that g is algebraic of degree at most M over $\mathbf{C}(f_1, \ldots, f_n)$. Since the characteristic is zero, it follows that the field of meromorphic functions on T is algebraic of degree at most M over $\mathbf{C}(f_1, \ldots, f_n)$.

Lemma 27. Let $\mathfrak{a}$ be a positive non-degenerate divisor on T and let $f_1, \ldots, f_m$ be a base for $L(3\mathfrak{a})$. Then the matrix of partial derivatives $\partial f_\mu/\partial z_\nu$ has rank n at any point not on the support of $\mathfrak{a}$, and the field $\mathbf{C}(f_1, \ldots, f_m)$ has transcendence degree at least n over $\mathbf{C}$.

Proof. We need only prove the first statement, since the second follows immediately from it. If it is false, let w be a point not on the support of $\mathfrak{a}$ at which the matrix has rank less than n. After a linear transformation on the z_ν, we may assume that

$$\partial f/\partial z_1 = 0 \text{ at } z = w \text{ for each } f \text{ in } L(3\mathfrak{a}). \tag{44}$$

Let $\theta(z)$ be a theta-function corresponding to $\mathfrak{a}$; then for each u, v in V

$$f(z) = \theta(z - u)\theta(z - v)\theta(z + u + v)/\theta^3(z) \tag{45}$$

is a multiply periodic function and is therefore in $L(3\mathfrak{a})$. Write $\phi = \theta^{-1}\partial\theta/\partial z_1$; then (44) applied to the $f(z)$ in (45) gives

$$\phi(w - u) + \phi(w - v) + \phi(w + u + v) - 3\phi(w) = 0.$$

This shows that $\phi(w - u) - \phi(w)$, considered as a function of u, is additive and therefore $\mathbf{Q}$-linear; and since it is meromorphic, it must even be $\mathbf{C}$-linear. To multiply $\theta(z)$ by the trivial theta-function $\exp\{2\pi i \Sigma\Sigma a_{\mu\nu} z_\mu z_\nu\}$ would increase $\phi(w - u) - \phi(w)$ by $-4\pi i \Sigma a_{1\nu} u_\nu$; so by replacing $\theta(z)$ by an equivalent theta-function we can arrange that $\phi(w - u) - \phi(w) = 0$. Integrating this with respect to u_1, we find that $\theta(u)\exp\{-u_1\phi(w)\}$ depends only on $u_2, \ldots, u_n$ and not on u_1; and this contradicts the non-degeneracy of $\mathfrak{a}$. This proves the Lemma.

Putting the last two Lemmas together we obtain

Theorem 28. Suppose that there is a non-degenerate positive divisor on T; then the field of meromorphic functions on T is finitely generated and of transcendence degree n over **C**.

In the general case, let W be the intersection of the kernels of all the positive semi-definite Hermitian Riemann forms on V; it is easy to see that there is a positive semi-definite Hermitian Riemann form on V whose kernel is precisely W, so that V/W and the associated lattice define a torus T' which satisfies the conditions of Theorem 28. It follows easily from Theorem 23 that every meromorphic function on T is induced by a meromorphic function on T', so the theory for T is essentially the same as the theory for T'. For a general Λ we saw in §5 that there are no non-trivial Riemann forms, so that W = V and the only meromorphic functions on T are constants.

Theorem 29. Let $\mathfrak{a}$ be a non-degenerate positive divisor on T; then $L(3\mathfrak{a})$ induces an embedding of T into projective space as a complete non-singular variety.

Proof. We can replace $\mathfrak{a}$ by any positive divisor linearly equivalent to it without altering the projective embedding; so by the Corollary to Theorem 24 we can assume that there is no non-zero translation which leaves $\mathfrak{a}$ fixed. In the notation of Lemma 27, the embedding takes z in V into the point

$$(f_1(z)\theta^3(z), \ldots, f_m(z)\theta^3(z)),$$

each of these coordinates being holomorphic on V. Since for every u, v in V,

$$\theta(z - u)\theta(z - v)\theta(z + u + v) \tag{46}$$

is a linear combination of these coordinates, there is no z in V at which they all vanish. To show that the map separates points it is enough to show that given any z' and z" in V which are incongruent mod Λ, we can choose u and v so that the function (46) vanishes at z' but not at z". For this we choose u so that z' - u is in the support of $\mathfrak{a}$ but z" - u is not, which is possible by the auxiliary hypothesis on $\mathfrak{a}$; and we then choose v in general position.

The image is non-singular at any point outside the support of $\mathfrak{a}$, by the first part of Lemma 27; and a similar argument works for the points on the support of $\mathfrak{a}$. The image lies in a complete variety of dimension n, by Theorem 28; since the image is compact and we have already seen that it is a complex manifold of dimension n, it must be the whole of the variety. This completes the proof of the Theorem.

Conversely if a positive divisor on T is degenerate, there are infinitesimal translations that leave it invariant; and every positive divisor linearly equivalent to it has the same Riemann form and therefore is invariant under the same infinitesimal translations. So the map into projective space induced by such a divisor does not even preserve dimension. Thus if T can be regarded as an algebraic variety at all, it must contain a non-degenerate positive divisor. Moreover the projective embedding of T given by Theorem 29 clearly carries with it the group structure on T; so the associated projective variety is an abelian variety in the sense of the Appendix. For this reason, we shall say that T is an <u>abelian manifold</u> (and will then usually denote it by A) if T contains a non-degenerate positive divisor.

Theorem 30. _Let_ $\mathfrak{a}$ _be a non-degenerate positive divisor on_ T _such that_ $L(\mathfrak{a})$ _induces an embedding of_ T _into projective space as a non-singular variety_ A; _and let_ d_0 _be the Pfaffian of the Riemann form associated with_ $\mathfrak{a}$. _Then_ A _has degree_ $n!\,d_0$ _and the ambient projective space has dimension_ $(d_0 - 1)$.

This Theorem is included for convenience of reference. The second statement is a special case of Theorem 24; but a rigorous proof of the first statement would be far outside the scope of this book. A sketch of a proof can be found in [2], pp. 206-10. The idea is to vary T and $\mathfrak{a}$ continuously until the result is reduced to the special case where T is a product of curves of genus 1 and the theta-function associated with $\mathfrak{a}$ can be written as the product of theta-functions associated with the curves. This special case follows immediately from the corresponding result for a curve of genus 1, which was proved in §5. A more natural proof could be obtained by cohomological methods.

III · Morphisms, polarizations and duality

7. Morphisms of Abelian manifolds and of tori

Up to this point we have considered a complex torus T as having an assigned structural map $\mathbf{C}^n = V \to T$ with kernel Λ. It is now necessary to show that T uniquely determines this structural map; in other words, to show that any diffeomorphism $T_1 \to T_2$ between two complex tori is essentially induced by a $\mathbf{C}$-vector space isomorphism $V_1 \to V_2$ which maps Λ_1 onto Λ_2. The key step is as follows:

Lemma 31. The holomorphic 1-forms on $T = V/\Lambda$ are just the $\Sigma\, a_\nu dz_\nu$ where the a_ν are constants.

Proof. At every point of T the z_ν form an admissible set of local coordinates. So any meromorphic 1-form on T can be written $\Sigma\, f_\nu(z) dz_\nu$ where the $f_\nu(z)$ are meromorphic on T; and the form is holomorphic if and only if all the $f_\nu(z)$ are holomorphic. But the only holomorphic functions on T are the constants, so the Lemma is proved.

Now on T the holomorphic 1-forms are a $\mathbf{C}$-vector space Ω_1 of dimension n; and we have a many-valued map $T \times \Omega_1 \to \mathbf{C}$ which takes $P \times \omega$ to $\int_O^P \omega$ where O is any pre-assigned point of T. The many-valuedness corresponds to the ability to alter $\int_O^P \omega$ by $\int_\Gamma \omega$ where Γ is a representative of any element of the homology group $H_1(T, \mathbf{Z})$, which is a free abelian group on 2n generators. If Ω_1^* denotes the dual space of Ω_1 there is a natural embedding of $H_1(T, \mathbf{Z})$ into Ω_1^*, given by mapping a closed curve Γ to the linear functional which takes ω to $\int_\Gamma \omega$; and this identifies Ω_1^* with V and the image of $H_1(T, \mathbf{Z})$ with Λ. Thus given only the complex manifold structure of T and a pre-assigned point O on it, we can recover V and Λ in a canonical way; and it is clear that a change of O merely corresponds to a translation on V. Moreover, this shows that a knowledge of the identity element on T and the complex manifold structure on T is enough to determine the group law on T.

Let T_1 and T_2 be complex tori with identity elements O_1 and O_2 respectively, and let $\phi : T_1 \to T_2$ be a holomorphic map; then we can write $\phi = \beta \circ \alpha$ where α, defined by $\alpha(P) = \phi(P) - \phi(O_1)$, is a holomorphic map which takes O_1 to O_2 and β is just translation by $\phi(O_1)$. So in considering maps $T_1 \to T_2$ we lose nothing by assuming that O_1 goes to O_2.

Theorem 32. Let T_1 and T_2 be complex tori with identity elements O_1 and O_2 respectively, and let $\phi : T_1 \to T_2$ be a holomorphic map such that $\phi(O_1) = O_2$. Then ϕ is a group homomorphism and, with the obvious notation, is induced by a **C**-linear map $\psi : V_1 \to V_2$ such that $\psi(\Lambda_1) \subset \Lambda_2$.

Proof. Let Ω_ν denote the **C**-linear space of holomorphic 1-forms on T_ν for $\nu = 1, 2$; then ϕ induces a linear map $\phi^* : \Omega_2 \to \Omega_1$ and hence also a dual map $\psi : \Omega_1^* \to \Omega_2^*$. If Γ is a curve on T_1, not necessarily closed, and ω is a 1-form on T_2, then $\int_\Gamma \phi^*\omega = \int_{\phi\Gamma}\omega$; if α_ν for $\nu = 1, 2$ denotes the canonical map which associates to a curve Γ on T_ν the element of Ω_ν^* which takes ω_ν in Ω_ν to $\int_\Gamma \omega_\nu$, it follows that $\psi \circ \alpha_1 = \alpha_2 \circ \phi$. In particular we can apply this to $H_1(T_1, \mathbf{Z})$. Since ϕ maps $H_1(T_1, \mathbf{Z})$ into $H_1(T_2, \mathbf{Z})$, and α_ν maps $H_1(T_\nu, \mathbf{Z})$ onto Λ_ν, this implies that $\psi(\Lambda_1) \subset \Lambda_2$, which completes the proof of the Theorem.

A particularly important sub-class of these maps ϕ consists of the isogenies. In the notation of Theorem 32, we say that ϕ is an isogeny if the underlying map $\psi : V_1 \to V_2$ is an isomorphism. If ϕ is an isogeny then its kernel is finite, ϕ is onto, and $\dim T_1 = \dim T_2$; conversely any two of these statements imply the third and also imply that ϕ is an isogeny. If we regard ψ as identifying V_1 with V_2 and call them both V, then $\Lambda_1 \subset \Lambda_2$ and we can consider ϕ as the canonical map $V/\Lambda_1 \to V/\Lambda_2$; the order of the kernel is $[\Lambda_2 : \Lambda_1]$ and we call this the degree of ϕ. In particular let δ be the identity map on T; then $m\delta : T \to T$ for any non-zero integer m is an isogeny which can be factorized as

$$V/\Lambda \to V/m^{-1}\Lambda \to V/\Lambda,$$

where the left-hand arrow is induced by the identity map on V and the right-hand arrow, which is an isomorphism, corresponds to multiplication by m. Clearly $m\delta$ has degree m^{2n} and its kernel is isomorphic to $(\mathbf{Z}/m\mathbf{Z})^{2n}$.

Lemma 33. Let $\phi_1 : T_1 \to T_2$ be an isogeny of degree m; then there is an isogeny $\phi_2 : T_2 \to T_1$ such that $\phi_2 \circ \phi_1 = m\delta$ on T_1 and $\phi_1 \circ \phi_2 = m\delta$ on T_2. Conversely let $\phi_1 : T_1 \to T_2$ and $\phi_2 : T_2 \to T_1$ be holomorphic homomorphisms and $m \neq 0$ an integer such that $\phi_2 \circ \phi_1 = m\delta$ on T_1 and $\phi_1 \circ \phi_2 = m\delta$ on T_2; then ϕ_1 and ϕ_2 are isogenies.

Proof. Suppose first that ϕ_1 is an isogeny of degree m, and as above write it in the form $V/\Lambda_1 \to V/\Lambda_2$ where the underlying map on V is the identity. Thus $\Lambda_1 \subset \Lambda_2$ and $[\Lambda_2 : \Lambda_1] = m$, whence $\Lambda_1 \supset m\Lambda_2$. The identity map on V induces

$$V/m\Lambda_1 \to V/m\Lambda_2 \to V/\Lambda_1 \to V/\Lambda_2$$

where the two outer arrows are ϕ_1 and the inner arrow defines an isogeny $\phi_2 : T_2 \to T_1$; by considering the composition of two successive arrows we obtain the results required.

For the converse, we note that $m\delta$ is onto and has finite kernel; so the equation $\phi_2 \circ \phi_1 = m\delta$ shows that ϕ_2 is onto and ϕ_1 has finite kernel. Using $\phi_1 \circ \phi_2 = m\delta$ similarly, we find that ϕ_1 and ϕ_2 are each onto and have finite kernel, and this is enough to show that they are both isogenies.

We shall say that T_1 is isogenous to T_2 if there exists an isogeny $T_1 \to T_2$. By Lemma 33 this relation is symmetric, and it follows easily that it is an equivalence relation. Now suppose that T_1 is an abelian manifold and T_2 is isogenous to T_1; then T_2 is also an abelian manifold. For we can take $T_1 = V/\Lambda_1$ and $T_2 = V/\Lambda_2$ with $\Lambda_1 \supset \Lambda_2$; now if H is a positive definite Hermitian Riemann form on V with respect to Λ_1, it is one also with respect to Λ_2.

There is an alternative way of defining an isogeny. Denote by

$\mathrm{Hom}(T_1, T_2)$ the additive group of holomorphic homomorphisms $T_1 \to T_2$ and by $\mathrm{End}(T)$ the ring of holomorphic endomorphisms of T; and write

$$\mathrm{Hom}_0(T_1, T_2) = \mathrm{Hom}(T_1, T_2) \otimes_{\mathbf{Z}} \mathbf{Q}, \quad \mathrm{End}_0(T) = \mathrm{End}(T) \otimes_{\mathbf{Z}} \mathbf{Q}. \qquad (47)$$

It follows from Lemma 33 that ϕ in $\mathrm{Hom}(T_1, T_2)$ is an isogeny if and only if it has both a left and a right inverse in $\mathrm{Hom}_0(T_2, T_1)$ and in this case the two inverses are the same. In particular $\phi : T \to T$ is an isogeny if and only if it is invertible in $\mathrm{End}_0(T)$. It will follow from Theorem 34 that if T is an abelian manifold this condition is equivalent to ϕ not being a divisor of zero; but if T is merely a complex torus this need not be so.

One major reason for the importance of isogenies is Poincaré's Complete Reducibility Theorem for abelian manifolds, various forms of which are stated below as Theorem 34 and its Corollaries; this too does not necessarily hold for complex tori.

Theorem 34. *Let* $\phi : A_1 \to A_2$ *be an element of* $\mathrm{Hom}(A_1, A_2)$ *and let* $A_3 \subset A_2$ *be the image of* ϕ. *Then* A_3 *is an abelian manifold*

$$\begin{array}{ccccc} A_1 & & \xrightarrow{\quad\phi\quad} & & A_2 \\ \downarrow & & & & \uparrow \\ A_3 \times B_1 & \longrightarrow & A_3 & \longrightarrow & A_3 \times B_2 \end{array}$$

and there is a commutative diagram as shown, in which B_1 *and* B_2 *are abelian manifolds, the vertical arrows are isogenies and the lower horizontal arrows are the obvious maps.*

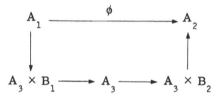

Proof. For $\nu = 1, 2$ write $A_\nu = V_\nu/\Lambda_\nu$ and $\dim V_\nu = n_\nu$; we shall extend this notation to $\nu = 3$ as soon as we have shown that A_3 is a complex torus. The map ϕ extends to a $\mathbf{C}$-linear map $V_1 \to V_2$ which we also denote by ϕ. Now $\phi\Lambda_1$ spans ϕV_1 as an $\mathbf{R}$-vector space, and it is discrete because it is a subgroup of Λ_2; so it is a lattice in ϕV_1 and $A_3 = \phi V_1/\phi\Lambda_1$ is a complex torus. If H_2 is a positive definite Hermitian Riemann form on V_2 with respect to Λ_2, its restriction to V_3 has the same properties with respect to Λ_3; so A_3 is an abelian manifold.

Now choose a positive definite Hermitian Riemann form H_1 on V_1 with respect to Λ_1; let W_1 be the kernel of ϕ acting on V_1, and let W_3 be the orthogonal complement of W_1 in V_1 with respect to H_1. Since Λ_1 and $\phi\Lambda_1$ have ranks $2n_1$ and $2n_3$ respectively, $\Lambda_1 \cap W_1$ must have rank $2n_1 - 2n_3$ and is therefore a lattice in W_1 since it is discrete. It is easy to see that W_3 is the orthogonal complement of W_1 with respect to E_1, the alternating form associated with H_1, and hence even the orthogonal complement of $W_1 \cap \Lambda_1$; since E_1 is integer-valued on $\Lambda_1 \times \Lambda_1$, this means that $W_3 \cap \Lambda_1$ has rank $2n_3$ and therefore it is a lattice in W_3. Let Λ_{10} be the projection of Λ_1 on W_1. Since the kernel of this projection is $W_3 \cap \Lambda_1$ which has rank $2n_3$, Λ_{10} has rank $2n_1 - 2n_3$; on the other hand it contains $W_1 \cap \Lambda_1$ which spans W_1 as an $\mathbf{R}$-vector space, so Λ_{10} spans W_1 and is therefore a lattice in W_1. Let B_1 be the complex torus W_1/Λ_{10}; then the projection $V_1 \rightarrow W_1$ induces a homomorphism $A_1 \rightarrow B_1$ and we have constructed the left-hand part of our diagram. Certainly $\dim A_1 = \dim A_3 \times B_1$. Moreover the projections $V_1 \rightarrow W_3$ and $\phi : V_1 \rightarrow V_3$ are both epimorphisms with kernel W_1; so they can be identified by means of a suitable isomorphism between W_3 and V_3. It is now clear that $A_1 \rightarrow A_3 \times B_1$ is onto, and therefore it is an isogeny. Finally, the proof that B_1 is an abelian manifold is analogous to that for A_3; it suffices to consider the restriction to W_1 of a suitable integral multiple of H_1.

The argument for the right-hand part of the diagram is similar but simpler. We have identified $V_3 = \phi V_1$ as a subspace of V_2; let W_2 be the orthogonal complement of V_3 in V_2 with respect to H_2. The argument that was used above for W_3 shows that $W_2 \cap \Lambda_2$ is a lattice in W_2; let B_2 be the complex torus $W_2/W_2 \cap \Lambda_3$, which is an abelian manifold for the same reason that A_3 is. The inclusion map $W_2 \subset V_2$ induces a homomorphism $B_2 \rightarrow A_2$, so the right-hand part of the diagram is constructed; and $\dim A_2 = \dim A_3 \times B_2$, so that the right-hand arrow, being onto, is an isogeny. This completes the proof of the Theorem.

Corollary 1. Any subtorus or quotient torus of an abelian manifold is an abelian manifold.

Proof. The argument which was used to prove A_3 an abelian manifold works for any subtorus of an abelian manifold. Now suppose that $A \to T$ is a homomorphism onto, where A is an abelian manifold and T a torus. The argument used to construct the left-hand side of the diagram in the Theorem shows that A is isogenous to $T \times T_1$ for some complex torus T_1; thus $T \times T_1$ is an abelian manifold and, by the part of the Corollary already proved, so is T.

To state the next Corollary, we need a further definition. We say that an abelian manifold A is simple if it is not isogenous to the product of two non-trivial complex tori (or of two non-trivial abelian manifolds, which comes to the same thing in view of Corollary 1). This is the same as saying that A has no non-trivial subtorus, or that it has no non-trivial quotient torus. Moreover, A is simple if and only if $\mathrm{End}(A)$ has no divisor of zero, for an element of $\mathrm{End}(A)$ is a divisor of zero if and only if its image has dimension lower than that of A. It now follows from Lemma 33 that A is simple if and only if $\mathrm{End}_0(A)$ is a division algebra.

Corollary 2. Any abelian manifold is isogenous to a product of simple abelian manifolds; and the factors are uniquely determined up to isogeny.

Proof. Induction on the dimension shows that any abelian manifold is isogenous to a product of simple abelian manifolds, so we have only to prove the uniqueness clause. Let

$$\phi : A_1 \times \dots \times A_m \to B_1 \times \dots \times B_n$$

be an isogeny, where the A_μ and B_ν are simple abelian manifolds. After re-ordering the B_ν if necessary, we can assume that $B_1 \times \dots \times B_r$ is the smallest partial product that contains ϕA_1. Now $\phi(A_2 \times \dots \times A_m)$ cannot contain $B_1 \times \dots \times B_r$, for otherwise it would contain ϕA_1 and so ϕ would be a dimension-reducing map. After further re-ordering we may assume that $\phi(A_2 \times \dots \times A_m)$ does not contain B_1; and therefore its intersection with B_1 is finite because B_1 is simple. Thus the map

$$\psi : A_2 \times \ldots \times A_m \to B_2 \times \ldots \times B_n,$$

which is the restriction of ϕ followed by projection, has finite kernel. Again, the projection of ϕA_1 on B_1 has image a subtorus of B_1; this cannot be just the identity element because then ϕA_1 would be contained in $B_2 \times \ldots \times B_r$, so it must be the whole of B_1 since B_1 is simple. The composite map $A_1 \to B_1$ is onto, and it has finite kernel because A_1 is simple, so it is an isogeny. Thus A_1 and B_1 have the same dimension, whence the source and target of ψ have the same dimension. So ψ is an isogeny too, and the Corollary now follows by induction.

This Corollary enables us to describe $\mathrm{Hom}_0(A, B)$ and $\mathrm{End}_0(A)$ for any abelian manifolds A and B. First note that if A and B are simple and not isogenous then $\mathrm{Hom}_0(A, B) = 0$. Next suppose that

$$A = A_1^{m_1} \times \ldots \times A_r^{m_r}, \quad B = A_1^{n_1} \times \ldots \times A_r^{n_r}$$

where $A_1, \ldots, A_r$ are simple and non-isogenous, and some of the m_i and n_i may vanish. Clearly $\mathrm{Hom}_0(A, B)$ is the direct sum of the $\mathrm{Hom}_0(A_i^{m_i}, A_i^{n_i})$, and this consists of the $n_i \times m_i$ matrices with elements in the division algebra $\mathrm{End}_0(A_i)$. The general case now follows by applying fixed isogenies to A and B; and a similar treatment works for $\mathrm{End}_0(A)$. However there is no such simple decomposition for Hom or End.

Corollary 3. Let A be an abelian manifold and B an abelian submanifold of A. Then there is an abelian submanifold C of A such that $B \cap C$ is a finite set and the natural map $B \times C \to A$ is an isogeny.

Proof. Apply the Theorem to the inclusion map $B \to A$, so that in the notation of the Theorem $A_2 = A$ and $A_1 = A_3 = B$; and let C denote the image of B_2 in A. The map $B_2 \to A$ has finite kernel, so $B \times C \to A$ is an isogeny; and P is in $B \cap C$ only if $(P, -P)$ is in the kernel of this isogeny. This proves the Corollary.

8. Duality of Abelian manifolds

Let $T = V/\Lambda$ be a complex torus; later in this section we shall need to assume that it is an abelian manifold. We saw in §5 that to any divisor on T corresponds a Riemann form. We shall say that two divisors on T are algebraically equivalent if the corresponding Riemann forms are the same; clearly this is an equivalence relation which is compatible with addition and subtraction of divisors, and it is weaker than linear equivalence. The terminology here comes from algebraic geometry, and the definition is an ad hoc one chosen to simplify the development of the theory; it would be more natural though less convenient to start from a different definition and prove the present one as a theorem.

The underlying idea is to consider two divisors as equivalent if one can be continuously transformed into the other within the set of all divisors. A Riemann form E with respect to Λ is determined by a matrix with integer elements; so the set of all possible Riemann forms with respect to Λ naturally has the discrete topology, and therefore two divisors which are equivalent in the sense of this paragraph must have the same Riemann form. Conversely if two divisors $\mathfrak{a}_1$ and $\mathfrak{a}_2$ are linearly equivalent then there are a divisor $\mathfrak{a}_0$ and a meromorphic function f on T such that $\mathfrak{a}_0 + \mathfrak{a}_1$ is the divisor defined by $f = 0$ and $\mathfrak{a}_0 + \mathfrak{a}_2$ by $f = \infty$. Thus the divisor

$$\{\text{divisor defined by } f = c\} - \mathfrak{a}_0$$

describes a continuous system as c varies, and the system contains both $\mathfrak{a}_1$ and $\mathfrak{a}_2$. This result can be extended to algebraic equivalence by means of Lemma 35 below, at least in the case when T is an abelian manifold; and the general case follows from this by means of Theorem 23 or the remarks after Theorem 28.

Lemma 35. Let A be an abelian manifold, $\mathfrak{a}$ a divisor on A algebraically equivalent to zero, and $\mathfrak{b}$ a positive non-degenerate divisor on A. Then there is a point x on A such that $\mathfrak{a}$ is linearly equivalent to $(\mathfrak{b} - \mathfrak{b}_x)$, where the suffix denotes translation by x.

Proof. By the argument leading up to Lemma 20, there is a meromorphic theta-function $\theta(z)$ associated with $\mathfrak{a}$ (that is, whose divisor is the pull-back of $\mathfrak{a}$ to V) with a functional equation of the form

$$\theta(z + \lambda) = \theta(z)\exp\{2\pi i f(\lambda)\}$$

for some homomorphism $f : \Lambda \to \mathbf{C}$. Let $\theta_1(z)$ be any theta-function associated with $\mathfrak{b}$, and suppose it has functional equation

$$\theta_1(z + \lambda) = \theta_1(z)\exp\{2\pi i[L(z, \lambda) + J(\lambda)]\}$$

where the conventions and notation are those of §5. Write

$$\psi(z) = \theta_1(z)\exp\{2\pi i R(z)\}/\theta_1(z - x)$$

where $R(z)$ is $\mathbf{C}$-linear; then $\psi(z)$ is a theta-function associated with $(\mathfrak{b} - \mathfrak{b}_x)$, and it satisfies the functional equation

$$\psi(z + \lambda) = \psi(z)\exp\{2\pi i[L(x, \lambda) + R(\lambda)]\}.$$

So to prove the Lemma we need only show that we can choose x and R so that

$$f(\lambda) = L(x, \lambda) + R(\lambda) \text{ for all } \lambda. \qquad (48)$$

The argument above Lemma 20 shows that we can take $L(z, w) = -\frac{1}{2}iH(z, w)$, and H is non-singular by hypothesis. In this case $L(x, w)$ is a $\mathbf{C}$-antilinear function of w, and can be made equal to any assigned $\mathbf{C}$-antilinear function by suitable choice of x. Again, f is the restriction to Λ of an $\mathbf{R}$-linear function on V, which can be written

$$f(z) = \tfrac{1}{2}\{f(z) + if(iz)\} + \tfrac{1}{2}\{f(z) - if(iz)\};$$

here the first term on the right is $\mathbf{C}$-antilinear and the second one is $\mathbf{C}$-linear. Thus (48) can be satisfied by suitable choice of x and R, and this proves the Lemma. It is not claimed (nor even true in general) that x is unique, for $f(\lambda)$ is really only determined mod $2\pi i$ and hence its extension $f(z)$ is not unique.

For any abelian manifold A, let G, G_a and G_l denote respectively the groups of all divisors, of divisors algebraically equivalent to zero, and of divisors linearly equivalent to zero; thus $G \supset G_a \supset G_l$. The group G/G_a, which is called the Néron-Severi group of A, is naturally isomorphic to the group of those Hermitian Riemann forms which can be written as the difference of two positive semi-definite Riemann forms; this is a free abelian group on at most n^2 generators, the precise number depending on number-theoretical properties of Λ. This group will not be further studied in this book. We now turn to the group G_a/G_l. To a divisor $\mathfrak{a}$ in G_a there corresponds an essentially unique normalized meromorphic theta-function $\theta(z)$, and the functional equation of $\theta(z)$ has the form

$$\theta(z + \lambda) = \theta(z)\exp\{2\pi i f(\lambda)\}$$

where $f : \Lambda \to \mathbf{R}/\mathbf{Z}$ is a homomorphism. Thus $\chi(\lambda) = \exp\{2\pi i f(\lambda)\}$ is a character on Λ, uniquely determined by $\mathfrak{a}$; and $\chi(\lambda)$ is trivial if and only if $\mathfrak{a}$ is in G_l. Conversely the construction of §6 shows that every character χ can be obtained from a meromorphic theta-function and therefore from a divisor $\mathfrak{a}$ - this is the step in the argument that would fail if A were merely a complex torus. We have therefore constructed a natural isomorphism of commutative groups

$$G_a/G_l \approx \text{Char } \Lambda . \tag{49}$$

But $\text{Char } \Lambda = (\mathbf{R}/\mathbf{Z})^{2n}$ is a real torus. Moreover, let $\mathfrak{b}$ be a fixed positive non-degenerate divisor; then by Lemma 35 the map which takes x in A to the class of $(\mathfrak{b} - \mathfrak{b}_x)$ is an epimorphism $A \to G_a/G_l$, and the proof of Lemma 35 implies that the composition of this map with (49) is continuous. It is therefore natural to hope that we can give Char Λ a complex structure in such a way as to make it an abelian manifold and to make the map $A \to \text{Char } \Lambda$ just described an isogeny.

For this purpose denote by $\hat{V}$ the space of $\mathbf{C}$-antilinear maps $V \to \mathbf{C}$ - that is, the space of $\mathbf{R}$-linear maps g such that $g(iz) = -ig(z)$; thus $\hat{V}$ is a $\mathbf{C}$-vector space of dimension n, as V is. Write $g_1(z) = \text{Im}\, g(z)$; then g_1 is an $\mathbf{R}$-linear map $V \to \mathbf{R}$ and

$g(z) = -g_1(iz) + ig_1(z)$. Conversely if g_1 is any **R**-linear map $V \to \mathbf{R}$ this formula defines a **C**-antilinear map g; so there is an isomorphism of **R**-vector spaces

$$\hat{V} \approx \{\mathbf{R}\text{-linear maps } V \to \mathbf{R}\}. \tag{50}$$

Denote by $\hat{\Lambda}$ the lattice in $\hat{V}$ which corresponds under this isomorphism to the dual of Λ; thus $\hat{\Lambda}$ is a lattice in $\hat{V}$ and consists of those g in $\hat{V}$ such that $g_1(\Lambda) \subset \mathbf{Z}$. The isomorphism (50) induces a map $\hat{V} \to \mathrm{Char}\,\Lambda$, which takes g in $\hat{V}$ to χ_g, where $\chi_g(\lambda) = \exp\{2\pi i g_1(\lambda)\}$; and the kernel of this is precisely $\hat{\Lambda}$. Thus we have defined an isomorphism of real tori

$$\hat{V}/\hat{\Lambda} \approx \mathrm{Char}\,\Lambda. \tag{51}$$

Henceforth we shall write $\hat{V}/\hat{\Lambda} = \hat{A}$, which is at any rate a complex torus since $\hat{V}$ is a **C**-vector space. Let $\mathfrak{b}$, as above, be a fixed positive non-degenerate divisor on A; and write

$$\phi : A \to \hat{A} \tag{52}$$

for the map $A \to G_a/G_l \to \mathrm{Char}\,\Lambda \to \hat{A}$, where the first arrow takes x in A to the class of $(\mathfrak{b} - \mathfrak{b}_x)$ and the other two arrows are the isomorphisms (49) and (51). From now on, ϕ will always denote this map; ϕ depends on $\mathfrak{b}$ but it is not necessary to make this dependence clear in the notation since we shall only consider one $\mathfrak{b}$ at a time.

Theorem 36. With the notation and hypotheses above, ϕ is an isogeny and $\hat{A}$ is an abelian manifold.

Proof. Let H and E be the Hermitian and alternating Riemann forms associated with $\mathfrak{b}$. In the notation of the proof of Lemma 35 we can take $L(z, w) = -\frac{1}{2}iH(z, w)$; and (48) then requires $R(z) = \frac{1}{2}iH(z, x)$ and $f(z) = E(x, z)$. The isomorphism (50) involves $g_1 = f$ and so finally $g(z) = H(x, z)$. This makes the map $V \to \hat{V}$ underlying ϕ explicit and shows that it is **C**-linear; a priori it was only **R**-linear. We already know that ϕ is onto, and A and $\hat{A}$ have the same dimension; so ϕ is an isogeny, and since A is an abelian manifold so is $\hat{A}$. This proves the Theorem. It also shows that ϕ really depends not on $\mathfrak{b}$ but only on H or E - that is, on the algebraic equivalence class of $\mathfrak{b}$. For this reason, it is often written ϕ_H in the literature.

We shall call $\hat{A}$ the <u>dual</u> of A. To make this terminology acceptable, we have to show that there is a natural identification of $\hat{\hat{A}}$ with A. For this purpose we identify z in V with the **C**-antilinear map on $\hat{V}$ defined by $z(g) = \overline{g(z)}$, where g runs through the elements of $\hat{V}$. It is easy to check that this definition is compatible with the structure of V as a **C**-vector space; and since the imaginary part of z(g) is $-g_1(z)$ in the notation above, the dual of $\hat{\Lambda}$ is Λ.

Let us temporarily denote the underlying pairing $V \times \hat{V} \to \mathbf{C}$ by

$$z \times g \to g(z) = \langle z, g \rangle_A$$

where the subscript will be omitted whenever possible. Given any element of Hom(A, B) or End(A), we shall use the same symbol also for the underlying map of vector spaces. Thus to any α in Hom(A, B) corresponds a **C**-linear map $\alpha : V_A \to V_B$, and thus also its transpose ${}^t\alpha : \hat{V}_B \to \hat{V}_A$ defined as usual by

$$\langle \alpha z, g \rangle_B = \langle z, {}^t\alpha g \rangle_A$$

for all z in V_A and g in $\hat{V}_B$; and ${}^t\alpha$ is **C**-linear. It is easy to check that ${}^t\alpha$ maps $\hat{\Lambda}_B$ into $\hat{\Lambda}_A$; so it induces a homomorphism ${}^t\alpha : \hat{B} \to \hat{A}$ which is called the <u>transpose</u> of α. Moreover, all the standard formalism works as one would hope, including the fact that ${}^{tt}\alpha = \alpha$ when we identify $\hat{\hat{A}}$ with A.

Let H be a given positive definite Hermitian Riemann form for A, and let ϕ be the isogeny (52) determined by H. We define the <u>Rosati anti-automorphism</u> on $\mathrm{End}_0(A)$ by

$$\alpha \to \alpha' = \phi^{-1}\,{}^t\alpha\phi. \tag{53}$$

Lemma 37. $H(\alpha z, w) = H(z, \alpha' w)$ <u>for any</u> z <u>and</u> w <u>in</u> V; <u>in other words</u> α' <u>is the adjoint of</u> α <u>with respect to</u> H. <u>Moreover</u> $\alpha'' = \alpha$, <u>so that (53) is an involution, and</u> ${}^t\phi = \phi$.

Proof. This is a straightforward calculation. It was shown in the proof of Theorem 36 that $\langle z, \phi w\rangle = H(w, z)$; so

$$H(\alpha' w, z) = H(\phi^{-1}\,{}^t\alpha\phi w, z) = \langle z, {}^t\alpha\phi w\rangle = \langle \alpha z, \phi w\rangle = H(w, \alpha z)$$

and on taking complex conjugates the first assertion follows. Hence also $\alpha'' = \alpha$. Again $H(z, w) = \langle w, \phi z\rangle = \langle {}^t\phi w, z\rangle$ whence $\langle z, \phi w\rangle = H(w, z) = \langle z, {}^t\phi w\rangle$ by taking complex conjugates; and this gives ${}^t\phi = \phi$ which completes the proof of the Lemma.

It would be nice if we could obtain from H a dual Riemann form $\hat{H}$ on $\hat{V}$ with respect to $\hat{\Lambda}$, in such a way that $\hat{\hat{H}}$ is the same as H; but this seems to be impossible. The obvious candidate would be

$$\hat{H}(f, g) = H(\phi^{-1}f, \phi^{-1}g), \tag{54}$$

but this is in general not a Riemann form because its imaginary part is not integer-valued on $\hat{\Lambda} \times \hat{\Lambda}$. However, a suitable multiple of $\hat{H}$ is a Riemann form, and we can maintain duality by introducing the idea of a polarization. A polarization P on an abelian manifold is a set P of positive non-degenerate divisors such that

(i) given any divisors $\mathfrak{a}_1$ and $\mathfrak{a}_2$ in P, there are non-zero integers n_1 and n_2 such that $n_1\mathfrak{a}_1$ is algebraically equivalent to $n_2\mathfrak{a}_2$; and

(ii) P is maximal subject to (i).

In analytic language, a polarization is just a positive definite Hermitian Riemann form H together with those of its multiples which are Riemann forms. The meaning of a map of polarized abelian manifolds is obvious, and if $A \to B$ is a map of abelian manifolds and B is polarized, there is an induced polarization on A.

For practical purposes one can afford to think of a polarization of A as equivalent to a projective embedding of A. Indeed by Theorem 29 a polarization induces a set of projective embeddings of A. If $\mathfrak{a}_1$ and $\mathfrak{a}_2$ are algebraically equivalent, then by Lemma 35 the projective embeddings they generate differ only by a translation. So the various projective embeddings induced by a given polarization are all related to one another by means of translations on the abelian manifold and Veronese mappings on the ambient space. Conversely any projective embedding of A determines a polarization.

It is easy to verify that the Rosati anti-automorphism (53) depends only on the polarization defined by H, and not on H itself. Moreover, if A is polarized then (54) determines a dual polarization on $\hat{A}$, with all the standard properties of duals; and ϕ is a map of polarized abelian varieties, and the analogous map $\hat{A} \to \hat{\hat{A}}$ is a multiple of ϕ^{-1}. Again, if α is in $\mathrm{End}_0(A)$ then straightforward calculation gives

$$^t(\alpha') = \phi\alpha\phi^{-1} = (^t\alpha)'.$$

If we use ϕ to identify V and $\hat{V}$, it follows from the proof of Theorem 36 that $\hat{\Lambda}$ is the dual of Λ with respect to the alternating Riemann form E; thus the degree of ϕ is equal to the square of the Pfaffian of E. We shall say that A is <u>principally polarized</u> if we can choose H in the polarization so that ϕ is an isomorphism; by the remarks above, this is the same as saying that by choosing a suitable base for Λ we can put the matrix of E in the form $\begin{pmatrix} 0 & I \\ -I & 0 \end{pmatrix}$. Not every abelian manifold admits a principal polarization; but by considering the matrix representation $\begin{pmatrix} 0 & D \\ -D & 0 \end{pmatrix}$ for E we easily see that any polarized abelian manifold is isogenous to a principally polarized one. In the more advanced theory, principally polarized abelian manifolds have a number of important properties which have not been proved and may not even be true for arbitrary abelian manifolds; see for example Maass [7] passim.

We defined the Jacobian of a compact connected Riemann surface S - or, which comes to the same thing, of an algebraic curve defined over **C** - in §1, just after the statement of Theorem 7. One reason why principal polarizations are important is the following result.

Theorem 38. <u>Let S be a compact connected Riemann surface; then its Jacobian is an abelian manifold which has a principal polarization.</u>

Proof. We use the notation of §1, and in particular of Theorem 8, so that g is the genus of S and $\Gamma_1, \ldots, \Gamma_g, \Gamma'_1, \ldots, \Gamma'_g$ is a normalized base for $H_1(S, \mathbf{Z})$. The map that takes ω to $(c_1, \ldots, c_g)$, in the notation of Theorem 8(ii), has trivial kernel by (8). So it is an isomorphism, by comparing dimensions, and we can choose a base

$\omega_1, \ldots, \omega_g$ for the differentials of the first kind on S such that $c_{\mu\nu}$, the period of ω_μ with respect to Γ_ν, is 1 if $\mu = \nu$ and 0 otherwise. Denote by $c'_{\mu\nu}$ the period of ω_μ with respect to Γ'_ν, and let Φ be the matrix of the $c'_{\mu\nu}$; then it follows from Theorem 8(i) that Φ is symmetric. Again, let $z_1, \ldots, z_g$ be any complex numbers not all zero, and write $\omega = \Sigma z_\mu \omega_\mu$ in Theorem 8(ii); then it follows from (8) that

$$\operatorname{Im} \Sigma\Sigma \bar{z}_\nu z_\mu c'_{\mu\nu} > 0,$$

whence $\operatorname{Im} \Phi$ is positive definite.

Now take the elements of Λ corresponding to $\Gamma_1, \ldots, \Gamma_g$, $\Gamma'_1, \ldots, \Gamma'_g$ as a base for Λ, and the first g of them as a base for the **C**-vector space V - which was called Ω^* in the notation of §1. Let H be the positive definite Hermitian form on $V \times V$ whose matrix representation with respect to this base is $(\operatorname{Im} \Phi)^{-1}$. Direct calculation shows that the matrix representation of E with respect to the given base of Λ is $\begin{pmatrix} 0 & -I \\ I & 0 \end{pmatrix}$; so H is a Riemann form and corresponds to a principal polarization. This proves the Theorem.

9. Representations of $\mathrm{End}_0(A)$

In what follows, $\mathbf{M}_m(F)$ will denote the ring of $m \times m$ matrices with elements in F, where $m > 0$ is an integer and F is any ring. There are two representations of End(A) and $\mathrm{End}_0(A)$ which are important in the analytic theory, and a third which can be described in the analytic theory but only becomes important in the geometric theory. Any element of $\mathrm{End}_0(A)$ induces an endomorphism of the underlying vector space $V = \mathbf{C}^n$; so there is a representation

$$\mathrm{End}_0(A) \rightarrow \mathrm{End}(\mathbf{C}^n) \approx \mathbf{M}_n(\mathbf{C})$$

which is called the <u>complex representation.</u> Again, any element of End(A) induces an endomorphism of Λ, so that there is a representation

$$\mathrm{End}(A) \rightarrow \mathrm{End}(\Lambda) \approx \mathrm{End}(\mathbf{Z}^{2n}) \approx \mathbf{M}_{2n}(\mathbf{Z});$$

and by tensoring with **Q** this induces

$$\mathrm{End}_0(A) \to \mathrm{End}(\Lambda \otimes \mathbf{Q}) \approx \mathrm{End}(\mathbf{Q}^{2n}) \approx \mathbf{M}_{2n}(\mathbf{Q}).$$

Either of these is called the rational representation. Obviously both the complex and the rational representation are faithful.

Now let l be a fixed rational prime; the letter l is used here by analogy with the geometric theory, in which p is reserved for the characteristic of the underlying field. For each $r > 0$, any element of End(A) induces a homomorphism on the group of l^r-division points of A (that is, on the kernel of $l^r\delta$); and this gives a representation

$$\mathrm{End}(A) \to \mathrm{End}((\mathbf{Z}/l^r)^{2n}) \approx \mathbf{M}_{2n}(\mathbf{Z}/l^r).$$

The sequence of representations corresponding to increasing r are compatible, and taking the limit we obtain a representation

$$\mathrm{End}(A) \to \mathrm{End}(\mathbf{Z}_l^{2n}) \approx \mathbf{M}_{2n}(\mathbf{Z}_l); \tag{55}$$

and once again tensoring with $\mathbf{Q}$ induces

$$\mathrm{End}_0(A) \to \mathrm{End}(\mathbf{Q}_l^{2n}) \approx \mathbf{M}_{2n}(\mathbf{Q}_l).$$

Either of these is called the l-adic representation. The group of l^r-division points can be identified with $l^{-r}\Lambda/\Lambda$, and the compatibility statement above corresponds to the homomorphism

$$l^{-r-1}\Lambda/\Lambda \to l^{-r}\Lambda/\Lambda \tag{56}$$

given by multiplication by l. Thus the $\mathbf{Z}_l$-module which determines the representation (55) is just the limit of the groups of l^r-division points under the homomorphisms (56). This is the Tate module $\mathbf{T}_l$, which is a free $\mathbf{Z}_l$-module of rank 2n; explicitly it can be described as the set of all infinite vectors $(a_1, a_2, \ldots)$ where a_r is an l^r-division point of A and $l a_{r+1} = a_r$ for each $r > 0$, and addition and multiplication are defined component-wise. The group of l^r-division points is also isomorphic to $\Lambda/l^r\Lambda$, and the homomorphism (56) is then replaced by the natural homomorphism $\Lambda/l^{r+1}\Lambda \to \Lambda/l^r\Lambda$; this makes the Tate module isomorphic to $\Lambda \otimes \mathbf{Z}_l$ and it follows that, so far as the analytic theory is concerned, the l-adic representation is equivalent to the rational representation and need not concern us further.

Once a coordinate system for V is chosen, any element of Λ can be written as an $n \times 1$ column matrix with complex elements. If we also choose a base for Λ, the complex torus V/Λ is completely described by means of the $n \times 2n$ matrix whose columns correspond to the elements of the chosen base for Λ; we shall denote this matrix by U. Varying the coordinate system in V corresponds to multiplying U on the left by an element of $GL(n, \mathbf{C})$; changing the base of Λ corresponds to multiplying it on the right by an element of $GL(2n, \mathbf{Z})$. We can use these operations to put U into canonical form; but when $V/\Lambda = A$ is an abelian manifold, which is the only case that interests us, it is convenient at the same time to make use of the existence of a polarization of A. So, as at the end of §5, we assume that the base for Λ is so chosen that the alternating Riemann form E associated with the polarization has the form

$$\begin{pmatrix} 0 & D \\ -D & 0 \end{pmatrix} \text{ where } D = \operatorname{diag}(d_1, \ldots, d_n)$$

and the d_ν are positive integers. As in the proof of Theorem 24, the last n elements of the chosen base for Λ are linearly independent over $\mathbf{C}$; so we can choose any convenient multiples of them as a base for V. We therefore suppose that the base for V is chosen so that U can be written

$$U = (S \;\; D) \text{ where } S = L + iM;$$

here S is in $\mathbf{M}_n(\mathbf{C})$ and L, M are in $\mathbf{M}_n(\mathbf{R})$. The condition that H defined by (37) should be Hermitian and positive definite is just that $E(ix, y)$ should be symmetric and positive definite; to express this, write

$$x = x_1 + ix_2 = (L + iM \;\; D) \begin{pmatrix} \xi_1 \\ \xi_2 \end{pmatrix}$$

where ξ_1, ξ_2 are real $n \times 1$ column matrices and x_1, x_2 are in the real vector space generated by $\lambda_{n+1}, \ldots, \lambda_{2n}$. Thus

$$\xi_1 = M^{-1}x_2, \quad \xi_2 = D^{-1}(x_1 - LM^{-1}x_2);$$

we already know that M is non-singular because the columns of U are linearly independent over $\mathbf{R}$. With a similar notation for y in terms of η_1 and η_2, we find

$$E(ix, y) = {}^t x_1 {}^t M^{-1} y_1 + {}^t x_2 M^{-1} y_2 + {}^t x_1 {}^t M^{-1}({}^t L - L)M^{-1} y_2.$$

This is symmetric and positive definite if and only if L and M are both symmetric and M is positive definite. The set of matrices $S = L + iM$ satisfying these conditions is called the <u>Siegel upper half-space</u>; it is one of the natural generalizations of the classical upper half-plane.

Given U, we can easily recover the images of the complex and the rational representations of $\mathrm{End}(A)$ and $\mathrm{End}_0(A)$. For let α be in $\mathrm{End}(A)$ and let γ and ρ be its complex and rational representations respectively; then $\gamma U = U\rho$. Conversely if γ in $\mathbf{M}_n(\mathbf{C})$ and ρ in $\mathbf{M}_{2n}(\mathbf{Z})$ are such that $\gamma U = U\rho$ then each of them defines an element α in $\mathrm{End}(A)$. For $\mathrm{End}_0(A)$ we need only replace $\mathbf{M}_{2n}(\mathbf{Z})$ by $\mathbf{M}_{2n}(\mathbf{Q})$.

Lemma 39. <u>The rational representation of $\mathrm{End}_0(A)$ is equivalent to the sum of the complex representation and its complex conjugate.</u>

Proof. The matrix $\binom{U}{\bar{U}}$ is non-singular because in the notation above its determinant is equal to $\det(2iM)\det(D)$ which is non-zero. Now let α be in $\mathrm{End}_0(A)$ and have rational and complex representations ρ and γ respectively; thus $\gamma U = U\rho$ and so $\bar{\gamma}\bar{U} = \bar{U}\rho$. These can be put together to give

$$\begin{pmatrix} \gamma & 0 \\ 0 & \bar{\gamma} \end{pmatrix}\binom{U}{\bar{U}} = \binom{U}{\bar{U}}\rho$$

and since $\binom{U}{\bar{U}}$ is non-singular this proves the Lemma.

Lemma 40. <u>Suppose that $\alpha \neq 0$ is in $\mathrm{End}_0(A)$ and α' is its image under the Rosati anti-automorphism; then $\mathrm{Tr}(\alpha\alpha') > 0$ where the trace is that derived from the rational representation.</u>

Proof. By Lemma 37, α' is adjoint to α with respect to the positive definite Hermitian form H; so by a standard theorem of elementary algebra the trace of the complex representation of $\alpha\alpha'$ is real and strictly positive. (Indeed, if we choose a base for V such that H is represented by the unit matrix, and denote the complex representation of α by the matrix (a_{ij}), then the complex representation of α' is $(\bar{a}_{ji})$ and the trace of the complex representation of $\alpha\alpha'$ is $\Sigma\Sigma|a_{ij}|^2 > 0$.) Lemma 40 now follows from Lemma 39.

The characteristic polynomial of the rational representation of an element α of $End_0(A)$ is called the rational characteristic polynomial of α; if we denote it by f_α then it has degree 2n and $f_\alpha(\alpha) = 0$ because the rational representation is faithful. There is an important alternative way of describing f_α. We can extend the definition of degree from End(A) to $End_0(A)$ by means of

$$\deg(r\alpha) = r^{2n}\deg(\alpha) \tag{57}$$

for any α in End(A) and r in **Q**. Now we have

$$f_\alpha(r) = \deg(r\delta - \alpha)$$

for any r in **Q** and α in $End_0(A)$, where δ is the identity; for when $r\delta - \alpha$ is in End(A) this merely states that the degree of an endomorphism is the determinant of its rational representation, which is obvious, and the general case follows from (57).

Using our results on the structure of U, we can now construct a classifying space for the set of abelian manifolds with a given polarization - that is, with a given matrix D. We have seen that to every point S of the Siegel half-space corresponds an abelian manifold with the given polarization, and that every such abelian manifold can be obtained in this way. However, a given abelian manifold corresponds to more than one S because there is more than one base for Λ which takes the matrix representation of E into canonical form. So if Σ denotes the Siegel upper half space and G the stabilizer of E in GL(2n, **Z**), the classifying space we are looking for is Σ/G where the action of G on Σ is still to be described. However, in order to remain compatible with the classical notation in the case $n = 1$ we proceed somewhat differently. Let σ be in GL(2n, **Z**); we wish σ to operate on Σ and therefore on U from the left, so we define its operation as $U \to U^t\sigma$, and we then have to multiply this on the left by a suitably chosen element of $GL(n, \mathbf{C})$ so as to renormalize it. Write

$$\sigma = \begin{pmatrix} \alpha & \beta \\ \gamma & \delta \end{pmatrix}; \tag{58}$$

then $U^t\sigma = (S^t\alpha + D^t\beta \;\; S^t\gamma + D^t\delta)$ and to renormalize we have to multiply on the left by $D(S^t\gamma + D^t\delta)^{-1}$. The resulting point of the Siegel space is therefore

$$D(S^t\gamma + D^t\delta)^{-1}(S^t\alpha + D^t\beta);$$

but we know this to be a symmetric matrix, so we can replace it by its transpose and write the transformation as

$$S \to (\alpha S + \beta D)(\gamma S + \delta D)^{-1}D.$$

Of course it would have been much more sensible to write U as a $2n \times n$ matrix ab initio, but in this matter we are the prisoners of history.

It remains to consider the action of σ on $E = \begin{pmatrix} 0 & D \\ -D & 0 \end{pmatrix}$, and with the conventions we have adopted this is clearly $E \to \sigma E^t\sigma$. So we define G, the symplectic group associated with D, to be the σ in $GL(2n, \mathbf{Z})$ such that

$$E = \sigma E^t\sigma. \tag{59}$$

This is equivalent to ${}^t\sigma E^{-1}\sigma = E^{-1}$; and in the special case $D = I$, which corresponds to the classical symplectic group, $E^{-1} = -E$ and so this condition becomes ${}^t\sigma E \sigma = E$. Using the partitioning (58) the condition (59) reduces to

$$\alpha D^t\delta - \beta D^t\gamma = D \quad \text{and} \quad \alpha D^t\beta,\ \gamma D^t\delta \ \text{symmetric.}$$

For this classification to be useful we need to show that G acts discontinuously on Σ - that is, that if N is a compact subset of Σ there are only finitely many σ in G such that $N \cap \sigma N$ is not empty. But if S is in $N \cap \sigma N$ the corresponding Hermitian form $H(z, z)$ is positive definite and lies in a compact set which depends only on N; so its value at any ${}^t\sigma^{-1}\lambda_\nu$, where $\lambda_1, \ldots, \lambda_{2n}$ is the given base for Λ, is bounded independently of σ. This means that the ${}^t\sigma^{-1}\lambda_\nu$ lie in a fixed finite set, and hence the same is true of σ. In particular it follows that a polarized abelian manifold only admits a finite group of automorphisms.

In contrast with the case $n = 1$, it is not known whether Σ/G naturally has the structure of a complex manifold. It seems likely that trouble can occur at fixed points of elements of G. Already in the case $n = 1$, Σ/G is not compact; but Baily and Borel [1] have shown for all n how it can be compactified in a satisfactory way, though with much greater difficulty than in the case $n = 1$.

10. The structure of $\mathrm{End}_0(A)$

We saw at the end of §7 that to determine the structure of $\mathrm{End}_0(A)$ for an arbitrary abelian manifold A it suffices to be able to do so when A is simple; moreover, if A is simple then $\mathrm{End}_0(A)$ is a division algebra. In this section we investigate the possibilities for $\mathrm{End}_0(A)$ for given $n = \dim A$. For the reader's convenience we begin by summarizing without proofs some standard results on division algebras and more generally on finite dimensional simple algebras; for proofs see Deuring [3] or Weil [18], Chapter IX. We know that $\mathrm{End}_0(A)$ is finite dimensional over $\mathbf{Q}$ because the rational representation is faithful.

Until further notice, let K be a field and F an algebra whose centre is K and which is finite dimensional over K; we also assume that F is <u>simple</u> over K - that is, F has no two-sided ideal other than $\{0\}$ and F itself. For any field $L \supset K$ we write $F_L = F \otimes_K L$; then F_L is a simple algebra with centre L and it is finite dimensional over L. The first major structure theorem is that any such F is isomorphic to a complete matrix ring $\mathbf{M}_n(D)$ where D is a division algebra; and F determines n uniquely and D up to isomorphism. Moreover, with the obvious identification of K with the centre of D, the dimension of D over K is a perfect square m^2; and every maximal subfield of D has degree m over K. We say that a field $L \supset K$ <u>splits</u> F if F_L is isomorphic to $M_r(L)$ for some r, which must be mn in the notation above; then L splits F if and only if there is a subfield of L which is

K-isomorphic to a maximal subfield of D. Every automorphism of F over K is inner - that is, is of the form $x \to c^{-1}xc$ for some invertible element c in F.

Now let D be a division algebra of dimension m^2 over its centre K and let α be any element of D. The characteristic polynomial of α, regarded as an endomorphism of the K-vector space D acting by multiplication on the left, is a perfect m^{th} power; its mth root, which has degree m, is called the reduced characteristic polynomial of α. Another way to define it is to take the minimal polynomial for α over K and choose that power of the minimal polynomial whose degree is m. The second and last coefficients of the reduced characteristic polynomial, with the standard sign changes, are called the reduced trace and the reduced norm of α over K. If k is a subfield of K with $[K : k]$ finite, we can define the reduced trace and the reduced norm of α over k by means of the usual tower rules; and they enjoy all the standard properties of traces and norms.

Suppose that F and G are simple algebras with centre K, both being finite dimensional over K; then $F \otimes_K G$ is a simple algebra with centre K, and the division algebra underlying it depends only on the division algebras underlying F and G. In this way we obtain a law of composition on the division algebras with centre K and finite dimensional over K. Under this law, these division algebras form a commutative group, called the Brauer group of K. The identity element of this group is K itself; the inverse of D is the division algebra which has the same elements as D but whose law of composition is defined by $d_1 \times d_2 \to d_2 d_1$. We write Br(K) for the Brauer group of K.

If K is an algebraic number field or a local field we can describe explicitly its Brauer group and the homomorphism $Br(K) \to Br(L)$ obtained by mapping the class of F, a simple algebra over K, to the class of F_L. We start with local fields; $Br(\mathbf{C})$ is trivial because there are no algebraic extensions of $\mathbf{C}$, and $Br(\mathbf{R})$ has order 2, the non-trivial element corresponding to the classical quaternions. Thus each of $Br(\mathbf{C})$ and $Br(\mathbf{R})$ has a unique embedding into $\mathbf{Q}/\mathbf{Z}$, which we shall use shortly. If K is an algebraic number field and $\mathfrak{P}$ a finite prime of K, then there is a canonical identification $Br(K_{\mathfrak{P}}) \approx \mathbf{Q}/\mathbf{Z}$. Now let K be an algebraic number field and F a simple algebra with centre K. For all but a finite set of primes $\mathfrak{P}$ in K, the algebra F splits over $K_{\mathfrak{P}}$; so writing $F_{\mathfrak{P}} = F \otimes_K K_{\mathfrak{P}}$, where the prime $\mathfrak{P}$ may be finite or infinite, the map that takes F to $\Pi F_{\mathfrak{P}}$ induces a homomorphism $\mathrm{Br}(K) \to \oplus \mathrm{Br}(K_{\mathfrak{P}})$. This can be embedded in an exact sequence

$$0 \to \mathrm{Br}(K) \to \oplus \mathrm{Br}(K_{\mathfrak{P}}) \to \mathbf{Q}/\mathbf{Z} \to 0,$$

where the penultimate map is given by addition in $\mathbf{Q}/\mathbf{Z}$ together with the canonical identifications above. Moreover if $L \supset K$ is another algebraic number field and $\mathfrak{q}$ in L is a prime lying above $\mathfrak{P}$ in K, then with these identifications the map $\mathrm{Br}(K_{\mathfrak{P}}) \to \mathrm{Br}(L_{\mathfrak{q}})$ is just multiplication by $[L_{\mathfrak{q}} : K_{\mathfrak{P}}]$. It follows that if F is a simple algebra with centre K, we can determine whether L splits F by examining the factorization in L of a certain finite set of primes in K, the set depending only on F. In particular, if the image of F in Br(K) has order r then there are extensions of K of degree r which split F, but none of lower degree; hence if D is the division algebra underlying F, the maximal subfields of D over K have degree r and so D has dimension r^2 over K. (This result need not be true if K is not an algebraic number field.) The particular case of this which we need is

Lemma 41. Let K be an algebraic number field and let D be a division algebra with centre K and finite dimensional over K. If D admits an anti-automorphism which leaves K elementwise fixed, then either D = K or D is a quaternion algebra over K.

Proof. The anti-automorphism can be regarded as an isomorphism between D and its inverse in the Brauer group; so the image of D in Br(K) has order 1 or 2. By the remarks above, this means that D has dimension 1 or 4 over K, and this proves the Lemma.

We now return to considering the structure of $\mathrm{End}_0(A)$, where A is a simple abelian manifold. Write $D = \mathrm{End}_0(A)$ so that D is a division algebra, and denote by K the centre of D; and choose once for all a polarization of A and its associated Rosati anti-automorphism, which is an involution by Lemma 37. Clearly this induces an involution on K;

we shall say that A is of the first kind if this involution on K is trivial, and of the second kind otherwise, and we shall write K_0 for the fixed field of the involution. Clearly $K = K_0$ in the first case and $[K : K_0] = 2$ in the second case.

Lemma 42. K_0 is totally real; and if A is of the second kind then K is totally complex and the involution induced on K is complex conjugacy.

Proof. Let α be any element of $D = \mathrm{End}_0(A)$. Since D contains no divisor of zero, the minimal polynomial for α over **Q** is irreducible and both its reduced characteristic polynomial (in the sense of this section) and its characteristic polynomial with respect to the rational representation described in §9 are powers of this minimal polynomial. Thus the corresponding traces are positive multiples of one another and Lemma 40 implies

$$\mathrm{Tr}_{D/\mathbf{Q}}(\alpha\alpha') > 0 \quad \text{if} \quad \alpha \neq 0; \tag{60}$$

if α is in K or K_0 this also applies to the trace for $K/\mathbf{Q}$ or $K_0/\mathbf{Q}$ respectively.

Suppose first that K_0 is not totally real and let σ_1 be a complex infinite prime of K_0 - that is, a complex embedding $K_0 \to \mathbf{C}$. We can find α in K_0 such that $\sigma_1\alpha$ is large and has argument near $\frac{1}{2}\pi$, while all its conjugates except $\overline{\sigma_1\alpha}$ are small. Thus $\mathrm{Tr}(\alpha\alpha') = \mathrm{Tr}(\alpha^2) < 0$ because its dominant term is $2\,\mathrm{Re}(\sigma_1\alpha)^2 < 0$, and this contradicts (60). So K_0 is totally real.

Now suppose that A is of the second kind and let σ_1 be a real infinite prime of K. Let σ be the prime of K_0 underlying σ_1 and let σ_2 be the other extension of σ to K; thus σ_2 is also totally real. We can find α in K so that $\sigma_1\alpha$ is large and positive, $\sigma_2\alpha$ is large and negative, and all the other conjugates of α are small. Thus $\mathrm{Tr}(\alpha\alpha') < 0$ because its dominant term is $2(\sigma_1\alpha)(\sigma_2\alpha)$, and this contradicts (60). So in this case K is totally complex; and since complex conjugacy induces an involution on K with fixed field K_0, it must be the Rosati involution on K. This completes the proof of the Lemma.

Recall that any quaternion algebra D with centre K has a canonical involution given by

$$x \to x^* = \mathrm{Tr}_{D/K}(x) - x$$

where the trace is the reduced one; the elements of D fixed by this involution are precisely those in K. Any K-involution of D is compounded of this and a K-automorphism of D, and therefore has the form $x \to c^{-1}x^*c$ for some non-zero c in D. This is certainly an anti-automorphism; it is an involution if and only if its square is the identity, and a little algebra shows that this happens if and only if $c^{-1}c^*$ commutes with every x in D. Thus $c^* = \lambda c$ with λ in K, and applying the canonical involution to this we see that $\lambda = 1$ or -1. In the first case c is in K and the involution is the canonical one; in the second case $\mathrm{Tr}(c) = 0$ and c^2 is in K.

With the notation above, we can give a complete account of the possibilities when A is of the first kind.

Theorem 43. Suppose that A is of the first kind, so that $K = K_0$ is totally real; then one of the three following cases holds:

(I) $D = K$.

(II) D is a quaternion algebra over K such that every component of $D \otimes_{\mathbf{Q}} \mathbf{R}$ is isomorphic to $\mathbf{M}_2(\mathbf{R})$; and there is an element c in D such that $c^* = -c$, c^2 is in K and totally negative, and the Rosati anti-automorphism is given by $\alpha' = c^{-1}\alpha^*c$.

(III) D is a quaternion algebra over K such that every component of $D \otimes_{\mathbf{Q}} \mathbf{R}$ is isomorphic to the classical quaternions; and $\alpha' = \alpha^*$.

Proof. Lemma 41 applied to the Rosati anti-automorphism shows that either $D = K$ or D is a quaternion algebra over K; and in the latter case the discussion above shows that the anti-automorphism must be of one of two kinds. To prove the Theorem we need only show that these two kinds of automorphism correspond to cases II and III. Note that $D \otimes \mathbf{R}$ is the direct sum of $r = [K : \mathbf{Q}]$ components, corresponding to the r embeddings of K into $\mathbf{R}$; each component is a simple algebra of dimension 4 over $\mathbf{R}$, and must therefore be isomorphic either to $\mathbf{M}_2(\mathbf{R})$ or to the classical quaternions. Moreover our two involutions on D induce corresponding involutions on each component; and we can show that $\mathrm{Tr}(\alpha\alpha') \geq 0$ for each component by replacing α by $b\alpha$, where b is in K and is large for one particular embedding of K into $\mathbf{R}$ and small for all the other embeddings.

Suppose first that $\alpha' = \alpha^*$. It is easy to check that $\mathrm{Tr}(xx^*)$ is non-negative for the classical quaternions, when it is equal to $2\,\mathrm{Norm}(x)$, but not for $\mathbf{M}_2(\mathbf{R})$, when it is equal to $2\,\mathrm{Det}(x)$; so every component of $D \otimes \mathbf{R}$ is isomorphic to the classical quaternions and we are in case III.

Now suppose that there exists c in D such that $c^* = -c$ and $\alpha' = c^{-1}\alpha^* c$. Choose a base 1, c, j, k for D over K such that c, j, k all anti-commute and their squares are in $\mathbf{K}$; then if $\alpha = a_1 + ca_2 + ja_3 + ka_4$ with the a_ν in K, calculation shows that

$$\mathrm{Tr}_{D/K}(\alpha\alpha') = 2(a_1^2 - c^2a_2^2 + j^2a_3^2 + k^2a_4^2).$$

Each component of this is non-negative if and only if c^2 is totally negative and j^2 and k^2 are totally positive; and the latter cannot happen if $D \otimes \mathbf{R}$ has a component isomorphic to the classical quaternions. So we are in case II and this completes the proof of the Theorem.

For completeness we list, as case IV, what happens when A is of the second kind. One can obtain further information about the structure of D in this case (see for example [8], pp. 196-200) but this involves using much deeper results about the structure of algebras.

(IV) D <u>is a division algebra whose centre</u> K <u>is a totally complex quadratic extension of a totally real field</u> K_0; <u>and the restriction of the Rosati anti-automorphism to</u> K <u>is complex conjugacy.</u>

Albert has shown that any division algebra of type I, II or III can be isomorphic to $\mathrm{End}_0(A)$ for a suitable A, and for type IV he has given necessary and sufficient conditions for this to be possible. However, the only easily answered question that one can ask is what constraints are imposed on $n = \dim A$ by fixing $\mathrm{End}_0(A)$. To consider this, write $e = [K : \mathbf{Q}]$, $e_0 = [K_0 : \mathbf{Q}]$ and $d^2 = [D : K]$.

Lemma 44. In case I, e divides n. In cases II and III, 2e divides n. In case IV, $e_0 d^2$ divides n.

Proof. Since D acts on $\Lambda \otimes \mathbf{Q} \approx \mathbf{Q}^{2n}$, the latter can be regarded as a vector space over D and therefore 2n is divisible by $[D : \mathbf{Q}] = ed^2$; this deals with cases II, III and IV. Now consider case I, and let E denote the alternating Riemann form associated with the given polarization of A. For any endomorphism α of A we have $\alpha' = \alpha$ because we are in case I, and it follows from Lemma 37 that $z \times w \to E(z, \alpha w)$ defines an alternating Riemann form on A. Its determinant is $\det(\Phi(\alpha))$ times that of E, where Φ denotes the rational representation; and since the determinant of any alternating Riemann form is a square, so is $\det(\Phi(\alpha))$. Apply this result to $m\delta - \alpha$, where δ is the identity endomorphism; it follows that

$$\det(\Phi(m\delta - \alpha)) = \det(mI - \Phi(\alpha))$$

is a square for every integer m. Since this is a polynomial in m of degree 2n, it must be the square of some polynomial f(m) with integer coefficients; and since f^2 is the characteristic polynomial of a faithful representation of α, this implies $f(\alpha) = 0$. Moreover f^2 and hence also f is a power of the minimal polynomial for α over $\mathbf{Q}$. This implies that $[\mathbf{Q}(\alpha) : \mathbf{Q}]$ divides n; choosing α so that it generates K over $\mathbf{Q}$ we find that e divides n, and this completes the proof of the Lemma.

Appendix · Geometric theory

Many of the results in Chapter III can be stated in purely geometric language, even though the proofs given there are analytic. The object of this appendix is to state how far they can be regarded as geometric theorems, by outlining without proofs some of the geometric theory of abelian varieties. The language used will be that of classical algebraic geometry; for an account in scheme-theoretic language see Mumford [8].

Let V be a variety, not necessarily complete, non-singular or irreducible. V is called a group variety if there is given a group law on the points of V such that the structure maps $V \times V \to V$ and $V \to V$ which respectively take $u \times v$ to uv and v to v^{-1} are regular maps in the sense of algebraic geometry. We say that a field k is a field of definition for V considered as a group variety if V, the two structure maps and the identity element of the group are all defined over k. In this case V must be non-singular, the irreducible component of V which contains the identity element is a normal subgroup of V and thereby a group variety defined over k, and the other irreducible components of V are cosets of this subgroup; for this reason we could without much loss have required V to be irreducible, and some writers do so.

An abelian variety over **C** should be so defined that it is just the geometric realization of an abelian manifold; in particular it should be complete and irreducible, and the group law should be commutative. It turns out that the first of these properties implies the last, so an abelian variety is defined to be any complete irreducible group variety.

Theorem 45. (i) The group law on an abelian variety is commutative.

(ii) Any abelian variety defined over **C** is also an abelian manifold; and conversely any projective embedding of an abelian manifold which is non-singular is an abelian variety.

The completeness of an abelian variety A also implies that any map from any variety V to A must be well-behaved in various ways:

Theorem 46. (i) Let $f : V \to A$ be a rational map of a variety to an abelian variety; then f is regular at every simple point of V.

(ii) Let $f : V \times W \to A$ be a rational map of a product to an abelian variety; then there are rational maps $f_1 : V \to A$ and $f_2 : W \to A$ such that $f(v \times w) = f_1(v) + f_2(w)$ for every simple point $v \times w$ on $V \times W$.

(iii) Let G be a group variety and $f : G \to A$ a rational map which takes the identity element of G to the identity element of A; then f is a homomorphism.

The last part is a generalized analogue of Theorem 32. One important consequence is that any map from the projective line to an abelian variety must be constant; for the projective line can be given a group structure corresponding to the additive group of the underlying field after removing one point, and a different group structure corresponding to the multiplicative group of the field after removing two points, and no non-constant map can be a homomorphism for both these group structures. Again, it follows that the group law on A is uniquely determined by a knowledge of the underlying variety and the identity element.

Theorem 47 (Chow). Let A be an abelian variety defined over k and let B be an abelian subvariety of A; then B is defined over some finite separable algebraic extension of k.

The main importance of this result is that it shows that there are no continuous families of abelian subvarieties of A - a result which is trivial in the analytic theory.

We can no longer use the definition of isogeny in §7; instead we say that a homomorphism $f : A_1 \to A_2$ is an isogeny if its kernel is finite, it is onto, and $\dim A_1 = \dim A_2$. As before, any two of these properties imply the third. The degree of f as an isogeny is its degree as a map; this need not be the same as the order of the kernel because of inseparability. With these conventions Lemma 33 remains true, and so isogeny is still an equivalence relation. Moreover the Poincaré

Complete Irreducibility Theorem 34 and Corollaries 2 and 3 of it remain true (with 'variety' for 'manifold' throughout, of course). Again, the map $m\delta$ still has degree m^{2n}; and in characteristic 0 or characteristic p with p prime to m, its kernel is still isomorphic to $(\mathbf{Z}/m\mathbf{Z})^{2n}$. However in characteristic p there is an integer ν depending only on A such that the kernel of $p^r\delta$ is isomorphic to $(\mathbf{Z}/p^r\mathbf{Z})^{\nu}$; ν satisfies $0 \leq \nu \leq n$ but can take any value in this range.

The standard definition of linear equivalence in algebraic geometry corresponds to the analytic definition given in §6; on the other hand, the definition of algebraic equivalence of two divisors on a torus given in §8 was purely ad hoc. However it can be shown that two divisors on an abelian variety defined over **C** are algebraically equivalent in the standard geometric sense if and only if the corresponding divisors on the associated abelian manifold are algebraically equivalent in the sense of §8 - indeed a proof of this is outlined in §8. A positive divisor on A is said to be non-degenerate if it is ample - that is, if some multiple of it induces a non-singular projective embedding of A; this corresponds to the definition in §6 in view of Theorem 29 and the remarks which follow its proof. With these new definitions Lemma 35 still holds; moreover we have the so-called Theorem of the Square:

Theorem 48. Let $\mathfrak{b}$ be any divisor on an abelian variety A and x, y any points on A. Then $(\mathfrak{b} - \mathfrak{b}_x - \mathfrak{b}_y + \mathfrak{b}_{x+y})$ is linearly equivalent to zero.

In the analytic theory this was too trivial to be worth stating; but in the geometric theory it is a difficult and important result. With the help of it, we can again give G_a/G_l the structure of an abelian variety (which we again call $\hat{A}$); and for any non-degenerate divisor $\mathfrak{b}$ on A we can again define an isogeny $\phi : A \to \hat{A}$ by mapping x on A to the point on $\hat{A}$ corresponding to the class of $(\mathfrak{b} - \mathfrak{b}_x)$. Moreover if A is defined over a field k then we can take $\hat{A}$ to be defined over k. All that part of the formalism of §8 which does not involve the Riemann form remains valid. The Néron-Severi group G/G_a is finitely generated, as indeed it is for any variety. A homomorphism of abelian varieties $\alpha : A \to B$ induces a homomorphism $\alpha^* : G(B) \to G(A)$ on the groups of divisors; and since this preserves both algebraic and linear equivalence it induces ${}^t\alpha : \hat{B} \to \hat{A}$. Moreover $\hat{\hat{A}}$ can be identified with A in such a way that ${}^{tt}\alpha = \alpha$ and ${}^t\phi = \phi$; and the Rosati anti-automorphism defined by (53) is an involution. The definition of a polarization is still valid in the geometric context; and we now say that a polarization of A is <u>principal</u> if we can choose a divisor $\mathfrak{b}$ in the polarization so that the associated ϕ is an isomorphism.

The Jacobian of a curve Γ is the group G_a/G_l realized as an abelian variety. This realization is obtained by means of a geometric construction which gives the Jacobian a natural projective embedding, or more precisely a natural polarization induced by the unique polarization on Γ. Theorem 38 can now be strengthened to say that this polarization is principal. Not every abelian variety is isogenous to a Jacobian, but the following result is almost as useful.

Theorem 49. Given an abelian variety A there is an abelian variety B (depending on A and not unique) such that $A \times B$ is isogenous to a Jacobian.

We now turn to the representations of $\mathrm{End}_0(A)$. In characteristic zero we can obtain a complex and a rational representation by embedding k, the least field of definition of A, into the complex numbers and regarding A as an abelian manifold. But even in this case the l-adic representations have certain advantages over the rational representation, although they are equivalent to it; for if $\bar{k}$ denotes the algebraic closure of k then the Galois group $\mathrm{Gal}(\bar{k}/k)$ acts non-trivially on the Tate module $\mathbf{T}_l$ and thus on the l-adic representation. The extra structure thus obtained is useful, particularly in number-theoretic applications.

In characteristic p it is known that neither the complex nor the rational representation can exist in general. Indeed for any p there is a curve Γ of genus 1, defined over the field of p^2 elements, for which $\mathrm{End}_0(\Gamma)$ is that quaternion algebra over $\mathbf{Q}$ which splits at all primes except p and infinity; and this algebra has no faithful representation in either $\mathbf{M}_1(\mathbf{C})$ or $\mathbf{M}_2(\mathbf{Q})$. However for $l \neq p$ the l-adic representation (55) is defined as in §9 and is faithful. Moreover, if we define ν by the statement that there are p^ν p-division points on A (so that $0 \leq \nu \leq n$)

then a similar construction gives a p-adic representation $\mathrm{End}(A) \to \mathbf{M}_\nu(\mathbf{Z}_p)$. This however is not necessarily faithful, though it certainly is so if $\nu = n$, or if $\nu > 0$ and A is simple.

Although in characteristic p we do not have a rational representation, we can still define the rational characteristic polynomial of an element of End(A). In accordance with standard conventions, if α is an endomorphism of A which is not an isogeny we write $\deg(\alpha) = 0$.

Lemma 50. Let α be any fixed element of End(A) and δ the identity element. There is a polynomial $f(X)$ of degree $2n$ and with coefficients in $\mathbf{Z}$ such that

$$f(m) = \deg(m\delta - \alpha)$$

for every integer m. Moreover $f(\alpha) = 0$ in End(A).

This can easily be extended to elements of $\mathrm{End}_0(A)$. In characteristic 0, $f(X)$ is clearly the characteristic polynomial of the rational representation of α; so in characteristic p we define it to be the rational characteristic polynomial of α and its second coefficient to be the trace of α. (Its constant term is just $\deg(\alpha)$ and needs no new name.) It turns out that this is also the characteristic polynomial of the l-adic representation of α, for $l \neq p$; moreover we can replace Lemma 40 by $\mathrm{Tr}(\alpha\alpha') > 0$ if $\alpha \neq 0$, with the new definition of trace.

To what extend can we describe the alternating Riemann form E, or at any rate the invariants $d_1, \ldots, d_n$ of its matrix representation, by purely geometric means? The first step is to mimic the pairing $\Lambda \times \hat{\Lambda} \to \mathbf{Z}$ induced by (50). Let $m > 0$ be an integer and $\hat{v}$ an m-division point on $\hat{A}$, and let $\mathfrak{a}$ be a divisor on A corresponding to $\hat{v}$; thus $m\mathfrak{a}$ is linearly equivalent to zero and is therefore the divisor of some function f. Moreover there is a function g on A such that $f(mx) = \{g(x)\}^m$ for all x on A. Now let u be an m-division point on A; then $g(x)$ and $g(x + u)$ have the same m^{th} power and so their quotient is an mth root of unity. It is easy to check that this depends only on u and $\hat{v}$, so we write

$$g(x + u) = e_m(u, \hat{v})g(x)$$

as a definition of the mth root of unity $e_m(u, \hat{v})$.

Lemma 51. In the notation above, e_m is a bilinear form; and in characteristic 0, or in characteristic p with m prime to p, it has trivial kernel in each argument.

Now fix a prime $l \neq p$ and let m run through the powers of l. For each $r > 0$ choose an isomorphism $\{l^r\text{-th roots of unity}\} \approx \mathbf{Z}/l^r$ in such a way that the diagrams

$$\begin{array}{ccc} \{l^{r+1}\text{-th roots of unity}\} & \to & \mathbf{Z}/l^{r+1} \\ \downarrow & & \downarrow \\ \{l^{r}\text{-th roots of unity}\} & \to & \mathbf{Z}/l^{r} \end{array}$$

commute, where the left-hand arrow is the lth power map. Then the limit of the functions e_m compounded with the isomorphisms above is a bilinear form

$$e : \mathbf{T}_l(A) \times \mathbf{T}_l(\hat{A}) \to \mathbf{Z}_l.$$

In characteristic 0 and with the natural choice of the isomorphisms above, this can be identified with the form $\Lambda \times \hat{\Lambda} \to \mathbf{Z}$ of (50).

Let $\mathfrak{b}$ be a divisor on A and ϕ the associated map $A \to \hat{A}$; this induces a map $\mathbf{T}_l(A) \to \mathbf{T}_l(\hat{A})$ which we also denote by ϕ. Then $\xi \times \eta \to e(\xi, \phi\eta)$ is a bilinear form $\mathbf{T}_l(A) \times \mathbf{T}_l(A) \to \mathbf{Z}_l$ which is easily shown to be alternating. In characteristic 0 it is just the l-adicization of E, bearing in mind that $\mathbf{T}_l(A)$ can then be identified with $\Lambda \otimes \mathbf{Z}_l$; thus from it we can read off the powers of l in the d_ν and by letting l run through all primes we can even recover the d_ν. Moreover in characteristic p these alternating forms give a partial substitute for the alternating Riemann form E.

Finally we consider the structure of $\mathrm{End}_0(A)$. Once again we can confine ourselves to the case when A is simple, so that we can use the notation of §10. Since, with the revised definition of trace, we still have $\mathrm{Tr}(\alpha\alpha') > 0$ for $\alpha \neq 0$, Theorem 43 and the classification of the possible $\mathrm{End}_0(A)$ into four types both remain valid and the arguments that lead up to them only require minor changes. However the proof of Lemma 44 depends essentially on the existence of the rational representation and the modification of the argument that is needed in characteristic p leads only to the following weaker result.

Lemma 52. _Let_ A _be a simple abelian variety in characteristic_ p, _and let_ e, e_0 _and_ d _be as in §10. Then in cases_ I _and_ III, e _divides_ n; _in case_ II, 2e _divides_ n; _and in case_ IV, $e_0 d$ _divides_ n.

For $n = 1$ this is best possible; in particular case III does occur. For $n > 1$ very little is known.

References

[1] Baily, W. L. and Borel, A. Compactification of arithmetic quotients of bounded symmetric domains. Ann. Math., 84 (1966), 442-528.

[2] Conforto, F. Abelsche Funktionen und algebraische Geometrie. Springer (1956).

[3] Deuring, M. Algebren. Berlin (1935).

[4] Gunning, R. C. Lectures on Riemann surfaces. Princeton (1966).

[5] Lang, S. Abelian varieties. Interscience (1958).

[6] Lang, S. Introduction to algebraic and abelian functions. Addison-Wesley (1972).

[7] Maass, H. Siegel's modular forms and Dirichlet series. Springer lecture notes, Vol. 216 (1971).

[8] Mumford, D. Abelian varieties. Oxford (1970).

[9] Shimura, G. and Taniyama, Y. Complex multiplication of abelian varieties and its applications to number theory. Publ. Math. Soc. Japan 6. Tokyo (1961).

[10] Shimura, G. Automorphic functions and number theory. Springer lecture notes, Vol. 54 (1968).

[11] Siegel, C. L. Topics in complex function theory, Vols. I to III. Wiley-Interscience (1972).

[12] Tannery, J. and Molk, J. Fonctions elliptiques. Paris (1893-1902).

[13] Weber, H. Lehrbuch der Algebra, Vol. III. Brunswick (1908).

[14] Weil, A. Sur les courbes algébriques et les variétés qui s'en déduisent. Paris (1948).

[15] Weil, A. Variétés abéliennes et courbes algébriques. Paris (1948).

[16] Weil, A. Théorèmes fondamentaux de la théorie des fonctions thêta. Sém. Bourbaki, 16 (1949).

[17] Weil, A. Introduction à l'étude des variétés kählériennes. Paris (1958).

[18] Weil, A. Basic number theory. Springer (1967).

Index of definitions